Galaxies in Turmoil

Galaxies in Turmoil

The Active and Starburst Galaxies and the Black Holes That Drive Them

Chris Kitchin

Chris Kitchin
University of Hertfordshire

Library of Congress Control Number: 2006936902

ISBN-10: 1-84628-670-0 e-ISBN-10: 1-84628-671-9
ISBN-13: 978-1-84628-670-4 e-ISBN-13: 978-1-84628-671-1

Printed on acid-free paper.

9 8 7 6 5 4 3 2 1

Springer Science+Business Media
springer.com

For Christine

Contents

Preface

Active galaxies involve some of the most extreme conditions and some of the most intriguing phenomena found anywhere in the universe and their study is amongst the hottest areas of research interest, yet there is currently no book that makes the topic available at a non-mathematical and not too technical a level. The purpose of this book is to try and fill this gap.

The book is aimed at readers who already have an interest in and some knowledge of astronomy, who wish to get to grips with the confusing plethora of types of active galaxies – perhaps in order to go on to study some aspect of the topic in greater detail, perhaps so that they can extend their observational program, perhaps simply because active galaxies are fascinating in their own right. The study of active galaxies dates back to the beginning of the twentieth century, but serious work on them has been undertaken only in the last five decades or so. New types of active galaxies and new aspects to existing known types have been found every time a new part of the spectrum – radio, infrared, ultraviolet, x-ray or gamma ray – has been opened up to observers. This has resulted in innumerable apparently different varieties of objects out there in extragalactic space and this can confuse and perhaps put off a student new to the subject. I hope that this book will go far towards reducing the muddle, and to this end Appendix 2 lists all the types of active galaxy names that I have been able to discover in current use, together with a brief note about each of them. I would recommend that the reader has frequent recourse to this appendix, at least until he or she has become familiar with the main classes of active galaxies.

Observers with small or moderate sized telescopes have probably been put off from trying to look at active galaxies since most of the images that they will have encountered will have come from the largest of professional telescopes. This impression though is false – some active galaxies can even be found using binoculars. Many more are accessible through the commonly available 0.15–0.3 m telescopes, and hundreds may be imaged if a CCD camera is also to hand. To try and encourage the observation and study of active galaxies by as wide a range of astronomers as possible, the final chapter of the book lists, among other things, the observing details of the brighter active galaxies.

Since the book is intended to provide a readable broad account of active galaxies, the use of equations has been (almost) completely avoided. Although active galaxies is a highly complex field of study, some of the very fine details have been omitted so that the reader can obtain an understanding of the whole subject without becoming too bogged down in the technicalities. Also the more specialist or secondary topics have been separated from the main text into boxes and these can be looked at by the reader or not as the need arises.

While the book is not intended to provide the basis for specialist or research level studies of active galaxies, it is possible that it may be useful as background material for anyone with such interests. In particular, since all aspects of all types of active galaxies are included, research workers may find it useful as a quick reference to the properties of and phenomena within those types of active galaxies that are outside their specialisms.

The ambiguous numbers "billion" and "trillion" are used here to mean 1,000,000,000 and 1,000,000,000,000 respectively. When it is needed, for example in converting redshifts to distances, the value of 71 km/s per megaparsec (21.5 km/s per million light years) has been employed. Finally I have used two terms throughout the book that may not be familiar to all readers. The first is the old-fashioned time period of an aeon. This equals 1,000,000,000 years and so is ideally suited to the discussion of the lives of galaxies. The second word is the megasun. This is a term that I have coined for a mass of a million solar masses and which is valuable during discussions of massive black holes and related topics.

I hope that you, the reader, will be as fascinated by active galaxies as I have been, and that some of the mystery and confusion that surrounds the topic will be reduced by this book. I hope also that some readers may be encouraged to go out with their binoculars and telescopes and see what's really happening out there for themselves.

Chris Kitchin, April 2006

1
Classical Galaxies

Summary

- How galaxies came to be recognized as large, independent and distant systems of stars, gas and dust outside the Milky Way Galaxy.
- The classification of galaxies based upon their visible shapes.
- How galaxies may have formed.
- The Milky Way Galaxy and its properties.
- How the expansion of the universe came to be discovered and the implications of that expansion.
- The age of the universe.

- Boxes
  Spectroscopy
  Doppler shifts
  The Big Bang
  Look-back time.

Readers already possessing a good background knowledge of astronomy may wish to proceed directly to Chap. 2; however unless they also have a good knowledge of spectroscopy, they are advised to read Boxes 1.1 and 1.2 before doing so.

1.1 The Great Debate

1.1.1 How It All Started

Many modern astronomers learn with some astonishment that people now in their early eighties were born before it was known that galaxies lay outside and far away from our own Milky Way and were themselves vast collections of stars and gas clouds. It is a measure of the stunning speed with which we have learnt about the universe that today's telescopes routinely study objects so far away that light has been traveling through space to reach us for nearly three times longer than the Earth has been in existence, whilst well under a century ago the best astronomers using the largest telescopes struggled to reach out a mere 2 million light years (0.6 Mpc). Of course, nebulae themselves could be observed further away than 2 Mly, but individual stars were another matter. Thus it was not until 1924 that Edwin Hubble obtained the first observational proof that one spiral nebula at least lay considerably beyond the limits of the Milky Way.

Hubble was trying to study individual stars within M 31, the great spiral nebula in Andromeda. Even with the largest telescope in the world, the 2.5-m (100-inch) Hooker telescope at Mount Wilson, Hubble could only detect the brightest single stars in M 31 (Fig. 1.1). He was hunting for the exploding stars called novae and in late 1923 thought that he had found one. A search through the Mount Wilson plate archive showed that the star had been imaged several times previously, going as far back as 1909. Armed with these observations, Hubble soon decided that he had found a Cepheid variable star and not a nova. To check this he observed again early in 1924. Sure enough, it was a Cepheid whose period was just over 31 days. This was an exciting result because a dozen years earlier Henrietta Leavitt, working at Harvard college observatory, had established that Cepheids' periods and absolute magnitudes[1] were related to each other. Thus Hubble was able to calculate his Cepheid's absolute magnitude to be about −3 and since its apparent magnitude averaged about 18.5, its distance must be some 700,000 ly (200 kpc). Now at that time there was some argument over the size of the Milky Way, but even the largest estimate was for a diameter of just 300,000 ly (100 kpc). M 31 must thus be at least 500,000 ly (150 kpc) beyond the Milky Way's most distant outskirts.

Hubble's results were published in 1925 and decisively ended the dispute about whether spiral nebulae were a part of the Milky Way or outside it, which had divided astronomers for decades. In fact we now

[1]Astronomers use the magnitude scale as a measure of the brightnesses of stars and other objects. Apparent magnitude is the brightness as the object actually appears in the sky, while absolute magnitude is the apparent magnitude that the object would have if its distance were 33 ly (10 pc). Differences between absolute magnitudes thus reflect real differences between the brightnesses of objects; differences between apparent magnitudes may be due to different distances to the objects involved and/or to differing actual brightnesses. A difference of one magnitude corresponds to a factor of ×2.512 (= $10^{0.4}$) between the brightnesses of two objects, a difference of two magnitudes is a factor of ×2.512^2 (= ×6.3), three magnitudes a factor of 2.512^3 (= ×15.9), etc. The scale is an inverse one, i.e. the brighter the object the lower the numerical value of the magnitude and the zero point is fixed so that stars of magnitude 6^m are just visible to the unaided eye from a good dark observing site. The apparent magnitudes of some sample objects are: Sun (-26.7^m), Sirius A (-1.5^m), Polaris ($+2.3^m$), Sirius B ($+8.7^m$), while the currently faintest detectable objects are about $+27^m$. The Sun's absolute magnitude for comparison is $+4.7^m$, while that of Sirius is $+1.4^m$.

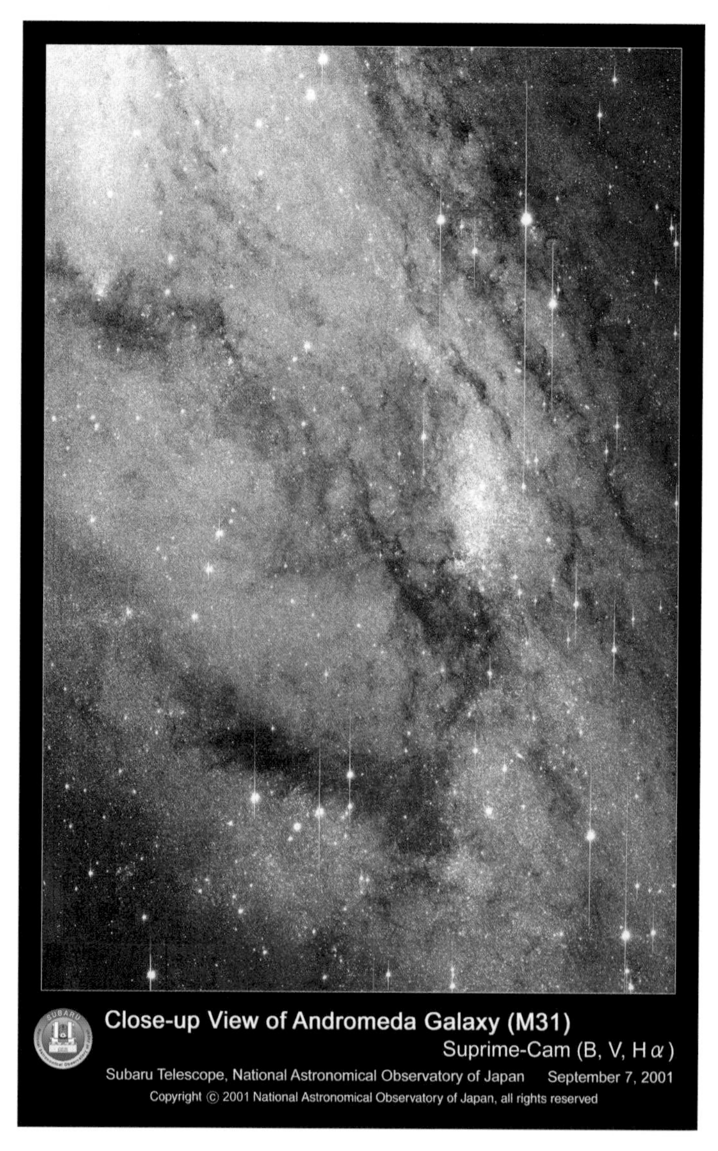

Figure 1.1 Individual stars within the Andromeda galaxy, M 31. Color image obtained using the Suprime-Cam on the 8-m (315-inch) Subaru telescope in 2001. (Copyright © Subaru Telescope, National Astronomical Observatory of Japan.)

know that the Milky Way galaxy (hereinafter called the Galaxy with an upper case G to distinguish it from all other galaxies) and the Andromeda galaxy are separated by almost 2 Mly (0.6 Mpc) and that the Galaxy is "only" about 100,000 ly (30 kpc) in diameter. Lord Rosse had first noticed spiral nebulae in 1845 whilst making visual observations using his 1.8-m (72-inch) reflector in Eire. The advent of photography enabled many more spiral nebulae to be found. The images also revealed other nebulae, some of which were irregular in shape while others had smooth, symmetrical, circular or elliptical shapes. These latter nebulae, even though bereft of any spiral structure, also became classed with the "spirals". By the end of the nineteenth century astronomers estimated the number of detectable spiral nebulae of both types to be in excess of 100,000.

Meanwhile, spectroscopy of nebulae undertaken both visually and photographically by William Huggins and others showed that in general they divided into two types; nebulae with bright emission lines like those that may be seen coming from hot gases in the laboratory, and nebulae with absorption lines like the spectra of stars (see Box 1.1). Generally the irregular nebulae had emission-line spectra, whilst the spirals had the absorption lines. Despite this evidence hinting that spiral nebulae were composed of stars, many astronomers, including Huggins, thought that the spiral nebulae were gaseous, perhaps even being planetary systems in the process of formation. Others scientists plumped for the "island universe" theory – the idea that spiral nebulae were composed of stars, maybe even millions of stars, but were so distant that they just appeared as blurs.

Box 1.1 Spectroscopy

The light that enables us to see things is just a small part of the complete electromagnetic (e-m) spectrum, which ranges from the longest radio waves to the shortest gamma waves. Confusingly e-m radiation sometimes behaves like a wave while at other times it behaves as though composed of particles. With wave-type behavior, we discuss e-m radiation in terms of its wavelength and frequency, while when it acts as though formed of particles; we talk about photons (or quanta) and their energies. The reasons for this apparently contradictory behavior lie deep within quantum theory and are beyond the scope of this book – the reader is simply asked to accept "that it is so" if not acquainted with that theory.

The dual nature of e-m radiation leads to wavelength and frequency tending to be used when discussing radiation in the ultraviolet, visual, infrared, microwave and radio regions, since interactions involving those types of e-m radiation tend to be dominated by wave-type behavior. While photon energies are apt to be invoked for x-rays and gamma rays since then the e-m radiation tends to have a particle-type behavior. To add to the confusion, a non-SI unit, the electron-volt (eV), is usually used as a measure for the energies of photons because the numbers then involved are more convenient. The electron-volt is defined as the energy gained by an electron when it is accelerated by one volt. Its value is $1\,\text{eV} = 1.6022 \times 10^{-19}\,\text{J}$. The divisions between regions of the e-m spectrum are arbitrary since its essential nature does not vary, but they are convenient and they are shown in Fig. 1.2 together with the generally accepted ranges of wavelength, frequency and photon energy. Additional sub-divisions are sometimes encountered such as soft and hard x-rays for the long wave and short wave x-rays respectively, extreme ultraviolet (EUV) for the shortest wave ultraviolet region, near, mid and far infrared (NIR, MIR, FIR), plus sub-millimeter waves for the region between the infrared and microwaves.

The complete e-m spectrum is infinite in extent, but it is customary to refer to a plot or image of a short segment of it that shows how intensity varies with wavelength, as a spectrum as well. Thus we have the rainbow where e-m radiation of different wavelengths within the visual part of the spectrum has been separated out by the effect of its passage through raindrops – although, strictly, this is not a pure spectrum since some wavelengths overlap each other. Purer visual spectra are obtained using instruments called spectroscopes. These employ diffraction gratings, interference filters, prisms, etc. to separate the various wavelengths (colors) so that they may then be imaged by a charge-coupled device (CCD) or photographic emulsion

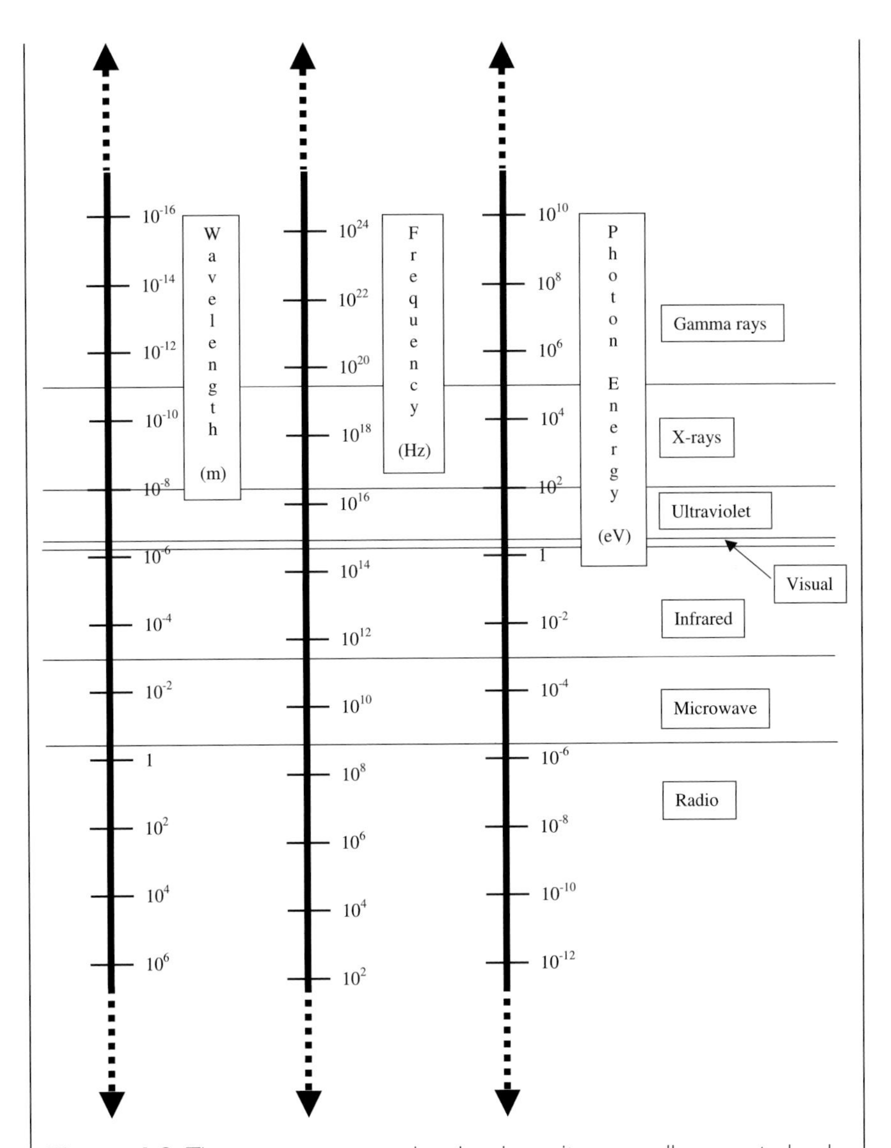

Figure 1.2 The e-m spectrum showing how its generally accepted subdivisions relate to wavelength, frequency and photon energy.

or seen directly by the eye, etc.[2] For other parts of the e-m spectrum, individual spectra may be produced using detectors that are intrinsically sensitive to different wavelengths, or which may be tuned to scan across a range of wavelengths. Other than for the visual region, spectra are usually shown as a plot of intensity versus wavelength, frequency or photon energy, and are then often called spectral energy distributions (SEDs), especially when the range covered is large.

Astronomers rely upon the spectra of the objects that they observe to provide a vast amount of information – probably more astronomical knowledge comes from spectroscopy than from all the other techniques used to study the universe put together. While not essential to acquiring an understanding of galaxies in general and active galaxies in particular, some further appreciation of what is involved with spectroscopy will help the reader considerably, especially in seeing *how* knowledge has been gained. Spectroscopy started in the visual region and since it is probably the most familiar part of the e-m spectrum to most people, we may use visual spectra to illustrate the concepts, nomenclatures and processes involved in studying galaxies spectroscopically, though identical or similar ideas are used whatever the wavelength involved.

Figure 1.3 illustrates the main types of spectra that will be encountered. At the top, resembling a rainbow, is the spectrum that would be observed from a hot solid, liquid or very dense gas and which is known as a continuous spectrum. Now, when they are in a gaseous form, the various chemical elements emit and absorb radiation at specific wavelengths in patterns unique to each element. Since light enters most spectroscopes through a slit and the spectrum is composed of images of the slit at each wavelength, the wavelengths

[2]There is not space here to go into further details of diffraction gratings, spectroscopes, etc. The interested reader may obtain further information from (amongst other sources) the author's book *Astrophysical Techniques*, 4 th edn, published by IoP Publishing, 2003.

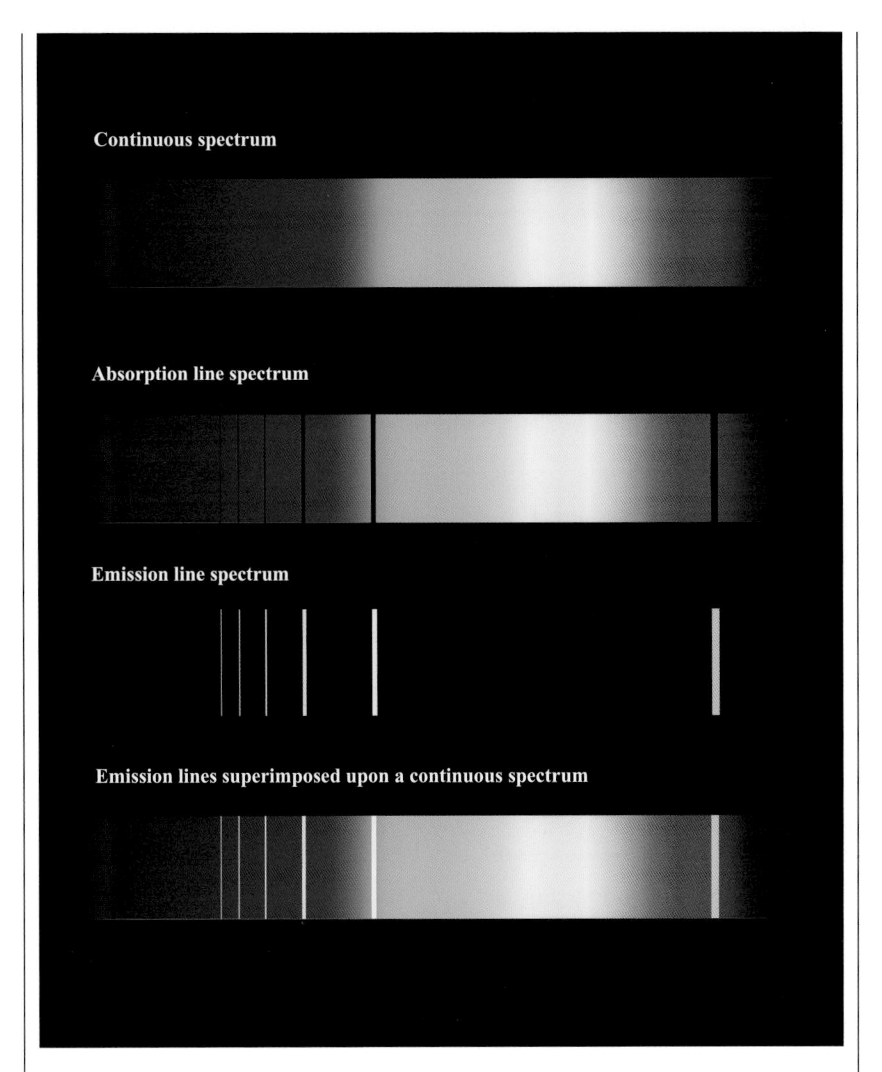

Figure 1.3 Types of visual spectra.

absorbed or emitted by an element appear as dark or bright lines running across the spectrum. These dark or bright regions are thus called the element's spectrum lines. At visible wavelengths, for example, hydrogen will normally have lines at wavelengths of 656.3 nm (red), 486.1 nm (blue–green), 434.0 nm (blue), and 410.2 nm (blue). If the density of the hydrogen gas is low then the series of lines may continue further towards the violet. Note that it is still quite common practice amongst some astronomers to use ångstroms as the unit for wavelength. Since 1 nm = 10 Å, the above wavelengths would become 6563 Å, 4861 Å, etc. and the reader should watch out for this if consulting other sources.

Pointing a telescope equipped with a spectroscope at stars and (most) galaxies will reveal spectra like the second one down in Fig. 1.3 and which is called an absorption-line spectrum. It is produced, to a first approximation, when the hot dense gas forming the inner layers of a star emits a continuous spectrum. That continuous emission then passes through the less dense outer layers of the star and the elements in those layers absorb at their characteristic wavelengths leaving the dark regions (absorption lines) in the final spectrum emitted from the star. Galaxies are seen by the light emitted by their millions of constituent stars, and so their spectra are composites of all the individual stellar spectra. The exceptions to this latter statement form the bulk of the material covered by the rest of this book and so will be left until later.

Hot gaseous nebulae such as H II regions (Box 2.2), planetary nebulae and supernova remnants produce spectra similar to the third example down in Fig 1.3. Here the thin gas forming the nebula is simply emitting at its characteristic wavelengths and so producing a pattern of bright lines against a dark background. Such a spectrum is called an emission-line spectrum (there may be a faint underlying continuous spectrum as well in many cases).

Finally the fourth spectrum in Fig. 1.3 is of the type to be found from many of the galaxies discussed later in this book and also from a few rare types of individual stars. It is an emission-line spectrum superimposed onto a continuous spectrum. It is also, somewhat confusingly, called an emission-line spectrum, although the context usually makes it clear whether it is the third or fourth type of spectrum shown in Fig. 1.3 that is intended. The latter type of emission-line spectra may well contain some absorption lines in addition to its emission lines.

Not only does each element have a unique pattern of spectrum lines, which enables the composition of even the most distant stars, nebulae and galaxies to be determined, but also if the atoms of an element lose an electron (become ionized) then another, different unique pattern of lines emerges. Losing second, third, fourth, etc. electrons (becoming doubly, triply, quadruply, etc. ionized) results in yet further different and unique patterns of lines. Rather more rarely an atom can gain an electron (become negatively ionized) and not surprisingly that also leads to a different pattern of lines. Since the level of ionization mostly depends upon the temperature of the gas, recognizing the patterns of ions' lines in spectra provides an estimate of the temperature of the region producing those lines. A common form of notation to indicate whether an element is present as neutral atoms or ions is the chemical symbol of the element followed by a Roman numeral whose value is one more than the number of missing electrons. Thus neutral iron is symbolized as Fe I, singly ionized iron as Fe II, doubly ionized iron as Fe III and so on. The rare negative ions are indicated with a negative sign as a trailing superscript, H^-, for negatively ionized hydrogen for example. Individual spectrum lines can now be labeled by their wavelength and state of ionization of the element. Thus the red line of hydrogen (Fig. 1.12, below) is H I 656.3.

Finally, in dealing with active galaxies later in this book we shall encounter "allowed" and "forbidden" spectrum lines (see also Box 3.1). All the lines so far mentioned have been "allowed". "Forbidden" lines are not actually forbidden – it is just that the probability of their occurrence is much lower under normal conditions than that for the allowed lines. Under normal conditions, such as may be found in most stars, forbidden lines are either undetectable or are very much weaker than the allowed lines of the same element. When the density of the gas is very low however, such as within gaseous nebulae and the interstellar medium, forbidden emission lines can become the strongest lines in the spectrum. Thus H II regions (Box 2.2) have emission-line spectra dominated by forbidden lines from ionized oxygen and nitrogen, while Seyfert galaxies' (Sect. 3.2.2) spectra contain strong emission lines from sulphur and neon in addition to those of oxygen and nitrogen. The nomenclature for allowed and forbidden lines is via the use of square brackets around the ion symbol. Both brackets are used when the line is strongly forbidden, but just a single bracket for lines whose probability of occurrence is somewhat stronger, but not as high as that of an allowed line. Thus [O III] 495.9 and [Ne III] 396.7 are forbidden lines, C II] 232.7 and Si III] 189.2 (both in the ultraviolet) are "semi"-forbidden lines, while He I 587.6 and H I 486.1 are allowed lines.

In April 1920 the leading proponents of the two theories, Harlow Shapley and Heber D. Curtis, delivered talks to the US National Academy of Sciences. The battle was called the "Great Debate" with Shapley putting the case for spiral nebulae being local and Curtis the case for their being distant galaxies. Their relatively brief lectures were followed by much fuller written presentations of their cases. However, despite the publicity that resulted from this clash, neither idea prevailed. So it was not until Hubble's discovery of Cepheids in M 31 four years later that the great debate was finally settled. Our modern view of the universe with our Milky Way Galaxy being just one out of millions of similar structures distributed over billions of light years, thus essentially dates from the publication of Hubble's 1925 paper.

1.1.2 Galaxies Today

Our present view of galaxies is that they are large collections of stars, brown dwarfs and planets, etc., together with varying proportions of gas and dust, which are held together as coherent entities by gravity. "Large" here means sizes ranging from 1,000 ly (300 pc) to 500,000 ly (150,000 pc) and containing between a million and a trillion stars. There are several characteristic shapes for galaxies including the well-known spirals (Sect. 1.2). The latter's splendor means that they dominate images of galaxies in books, etc. (including this one), but they are not actually the commonest type.

Galaxies may usually be distinguished from globular clusters because the latter, although overlapping with galaxies at the large end of their range, are spherical and generally composed of very old stars that were born nearly simultaneously. Most globular clusters are also satellites of larger galaxies, especially the giant elliptical galaxies (Sect. 1.2). Where galaxies have recently undergone collisions and mergers, globular clusters may be found that are composed of younger stars, so the distinction between them and galaxies is more difficult.

Galaxies may exist within the universe in isolation or by linked be gravity to other galaxies forming clusters. Clusters of galaxies may contain thousands of individual galaxies like the Virgo cluster or just a few like the local group to which the Milky Way Galaxy belongs. There are also clusters of clusters of galaxies with tens of thousands of members – the local supercluster, which includes the Milky Way, is centered on the Virgo cluster and stretches over some 100 Mly (30 Mpc). Whether or not we then get clusters of superclusters is still somewhat uncertain, although the consensus is currently that the hierarchy stops at superclusters.

1.2 Galaxies to the Forefront

1.2.1 Sorting Out the Galaxies

Hubble's work meant that the known volume of the universe increased at least 10-fold overnight. Indeed, taking the modern values for the size of the Galaxy and the distance to M 31, the increase in volume was actually by a factor of 8,000, and Hubble soon went on to observe galaxies much further out than M 31. Today, the volume of the visible universe is known to be some 2,000 trillion times that of the Milky Way Galaxy, and around a trillion other galaxies are to be found within that region.

Hubble's access to the Hooker telescope meant that few other astronomers could compete with him when it came to studying galaxies. He made the most of his opportunities and the year following the publication of his M 31 Cepheids paper, he had accumulated sufficient observations to start sorting out a classification system for galaxies. Classification does not sound all that exciting, yet it is one of the most powerful of scientific tools. It is almost always one of the first courses of action for researchers whenever a new field of study opens up in science. Classification serves to highlight and identify those objects that are normal and representative and to pick out the ones that are atypical. As the field of study matures and the regular objects begin to be understood, it is then the atypical ones that become the most interesting and informative. Several more sophisticated galaxy classification schemes have been developed since Hubble's work, but his original, relatively simple, system is still the most widely used and unless very specialized work is being undertaken, it is the most useful.

The classification scheme proposed by Hubble divides the galaxies into three main morphological types – spiral, elliptical and irregular. It then goes on to sub-divide these main groupings. The spiral galaxies, which had originally inspired the Great Debate, are first separated into "normal" and "barred" forms. A barred spiral galaxy has its spiral arms arising, not directly from the galaxy's nucleus, but from the ends of a rectangular "bar" that projects outwards from opposite sides of the nucleus (Figs. 1.4 and 1.5), while for a normal spiral galaxy the arms arise directly from the nucleus (Figs. 1.4 and 1.5). The normal and barred types are then further divided into three on the basis of the relatives sizes of their nuclei and the openness of the spiral winding of the arms. Sa and SBa (the "B" is for "barred") types thus have tightly wound arms and large nuclei, while Sc and SBc have much less tightly wound arms with a clumpy appearance due to the presence of many H

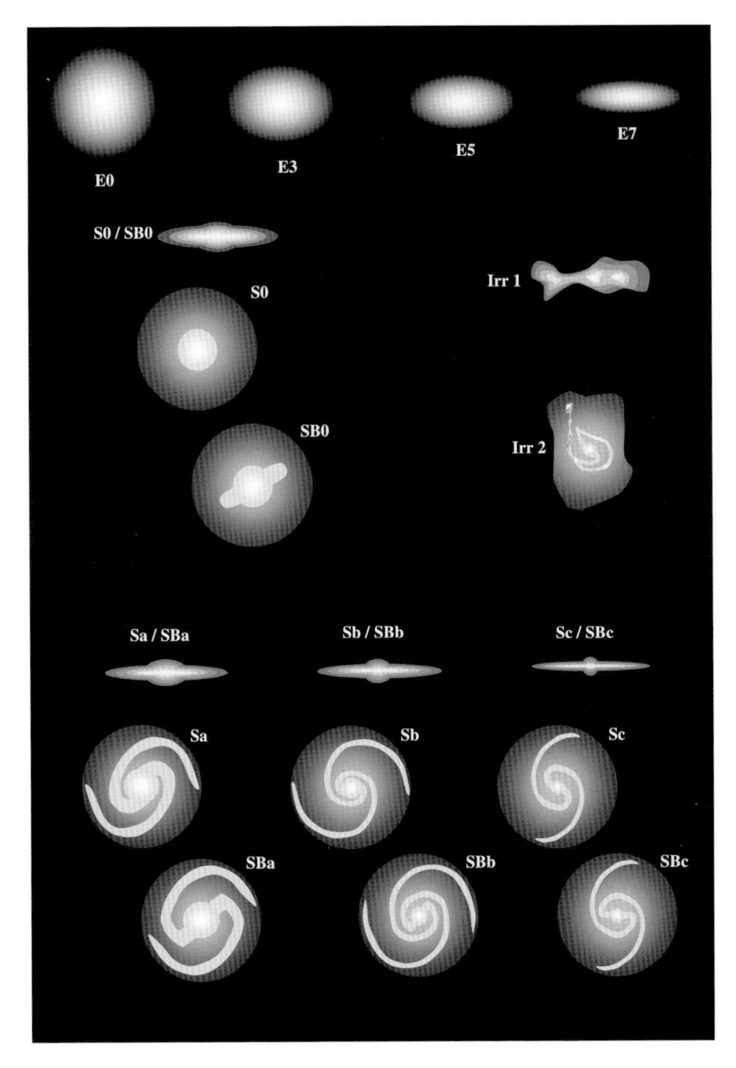

Figure 1.4 The Hubble system of galaxy classification. The galaxies are shown with their longest dimensions to the same scale, but in practice galaxies of all types may be found over a range of sizes (Table 1.1). Irregular galaxies are often small whilst giant ellipticals make the largest known galaxies. The spiral galaxies are shown with just two arms, but there may be more, and with the flocculent spiral galaxies there are many short segments of arms.

II regions (Box 2.2) together with relatively small nuclei. Sb and SBb types occupy the intermediate position. About half of the spiral galaxies clearly have a central bar, though weaker and hidden bars may well be present in most "normal" spiral galaxies.

Figure 1.5 Hubble classes of spiral galaxies:
(i) M 63, the Sunflower galaxy, an Sb type galaxy. A combination of two 5-minute exposures with a 0.4-m Schmidt–Cassegrain telescope, imaged using a STL1301E CCD camera. Log-stretching is used to display the image to its best advantage in print form. (Image courtesy of Bob Forrest and the University of Hertfordshire Observatory.)
(ii) An HST image of NGC 3370, an Sc type galaxy. (Image courtesy of STScI.)

Figure 1.5 *Continued*
(iii) M 33, an Scd type galaxy. A combination of two 5-minute exposures with a 0.4-m Schmidt–Cassegrain telescope, imaged using a STL1301E CCD camera. Log-stretching is used to display the image to its best advantage in print form. (Image courtesy of Bob Forrest and the University of Hertfordshire Observatory.)
(iv) An HST image of NGC 1300, an SBb type galaxy. (Image courtesy of NASA, ESA and the Hubble heritage team (STScI/AURA).)

Later workers, especially Gérard de Vaucouleurs, have added to Hubble's scheme, most notably by identifying classes Sd and SBd that have even greater degrees of openness and by adding a luminosity classification. De Vaucouleurs also labeled normal galaxies as types SA to match the SB of the barred galaxies, so that the reader may come across classes such as SAc that are identical to Hubble's class Sc. Here, since we are concerned mostly with galaxies that are not of a standard form, Hubble's original scheme will suffice.

Spiral galaxies may be divided independently from Hubble's system into "Grand Design" and "Flocculent" types on the basis of the appearance of their spiral arms. Grand design spirals, such as NGC 1300 (Fig. 1.5), have well-defined symmetrical and lengthy arms. Usually there are just two arms, but three or four are also possible. Flocculent galaxies have a spiral appearance, but when examined more closely this arises from numerous short segments of "arms" distributed chaotically. The differences reflect the different mechanisms underlying the formation of the spiral patterns. A grand design galaxy's arms are thought to develop as the result of a density wave propagating through its disk. The wave increases the density of the disk material, leading to increased rates of star formation. The arms then become visible through the high luminosities of massive young stars. Since such stars have short lives, the arms fade away at the trailing edge of the density wave as the massive stars die while new young stars are added at the leading edge. In fact, many less massive, longer-lived stars will be left behind as the arm moves on, so that the spiral appearance is deceptive. With the flocculent spiral galaxies, their short spiral segments are star-forming regions that have been dragged into their shapes during their orbital motion around the galaxy. Objects' spatial velocities within the disk of a galaxy do not fall off with distance in the Keplerian fashion of the solar system's planets, but are more or less constant. Nonetheless, material further from the nucleus has a greater distance to travel to complete its orbit, and so the angular velocity decreases outwards through the disk. Thus the outer parts of star-forming regions will gradually fall behind the inner parts, elongat-

Figure 1.6 NGC 221 (M 32), an E2 type galaxy that is also sometimes classified as a peculiar galaxy. (Image courtesy of Tod Lauer.)

ing the regions and so leading to the spiral appearances of flocculent galaxies.

Elliptical galaxies have no obviously visible internal structure; they are simply elliptical in shape with a sharp drop in brightness from their centers to their edges. Hubble therefore just divided them up on the basis of their degree of ellipticity. He identified eight classes from E0 to E7. The number for the class is related to the ellipticity – E0 galaxies appear circular, while E 7 galaxies have an aspect ratio[3] of 3.3 (Figs 1.4 and 1.6). Hubble's classification however is of the apparent shapes of the galaxies, i.e. as they are seen projected against the sky and does not

[3]The aspect ratio is the length of an object divided by its width – the larger the aspect ratio, the longer and thinner is the object. The number (N) for Hubble's elliptical galaxy classification is related to the galaxy's length (L) and width (W) by $N = \frac{10(L-W)}{L}$.

take into account their inclinations to the line of sight. The galaxies' true shapes are thus not reflected by the classification scheme. An ellipsoidal galaxy (one that has different diameters along all three axes) will always appear elliptical in shape, while a spheroidal galaxy (a sphere squashed along one axis, like a discus) can seem circular in shape if observed perpendicularly to its disk (Fig. 1.7).

Hubble included a further structured type of galaxy within his system – the lenticular galaxy, designated S0 or SB0. These may be pictured as spiral galaxies without spiral arms. They have a nucleus, sometimes relatively large and sometimes drawn out into a bar, surrounded by a disk, but the disk is featureless.

Within rich (large and densely populated) clusters of galaxies about 40% of the brighter galaxies are elliptical in shape, but outside such clusters the proportion is closer 10%. Some 80% of the bright

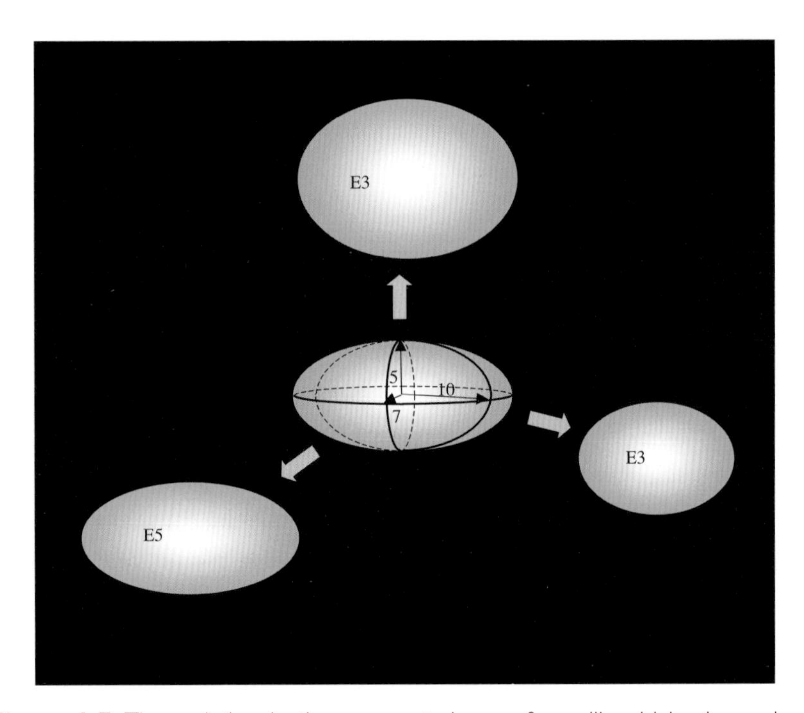

Figure 1.7 The variation in the apparent shape of an ellipsoidal galaxy whose dimensions are in the proportions 10:7:5 when it is observed from three mutually perpendicular directions.

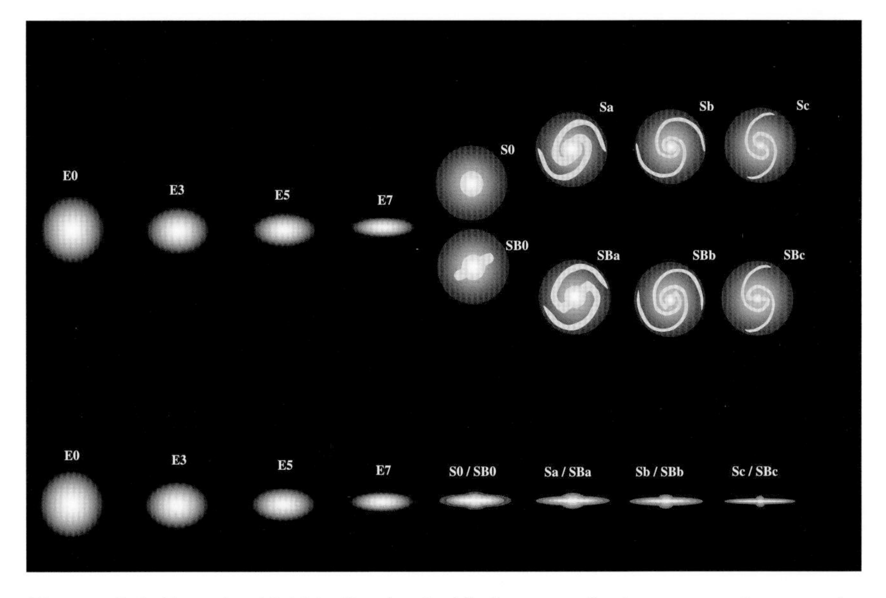

Figure 1.8 Top: the Hubble "tuning fork" diagram of galaxy types. Bottom: the same diagram with the galaxies seen from the side.

galaxies outside rich clusters are spiral in shape. Amongst galaxies of all luminosities, whether in clusters or not, ellipticals amount to about 60% and spirals to 30% of the total.

These various classes of galaxy are sometimes represented as a sequence, starting at E0 and ending with Sc and SBc, the two lines of spiral galaxies being separated to give the appearance of a "tuning fork" (Fig. 1.8). Hubble and others, at one time regarded this as an evolutionary[4] sequence, so that elliptical galaxies are occasionally called "early" type galaxies, and spirals "late" type. This idea is no longer accepted although some inter-type transformations probably do occur – two spirals for example colliding and merging to produce a giant elliptical galaxy. As usually represented, the tuning fork diagram does not seem

[4]Astronomers tend to use the word "evolve" to mean the changes and development of a single entity with time (i.e. its life cycle), not in the biological sense of a change in the nature of a large population of such entities arising from mutations.

to be a logical sequence of morphological types (Fig 1.8 top), but if the galaxies are seen from the side, then it is revealed to be a succession with a uniformly increasing aspect ratio (Fig. 1.8 bottom).

Unsurprisingly some galaxies did not fit neatly into any of Hubble's groups and so he added the Irregular class to act as a dustbin for all this galactic bric-a-brac. However even irregular galaxies can have some structure and common features and so the group had to be split into the Irr 1 and Irr 2 sub-divisions (Fig. 1.4). Irr 1 galaxies are genuinely irregular. They have a high gas content and contain star-forming regions that show up as bright patches within the galaxy. The bright patches are usually distributed unevenly and without any overall pattern. Though for some Irr 1 galaxies, a hint of a spiral structure *can* be found, so perhaps they represent the most extreme form possible for spiral galaxies.

Irr 2 galaxies do not fit into any of the other categories and often contain large amounts of dust, or have distorted shapes. Many of them show signs, such as bright star-forming regions or tidal deformations, suggesting that they are galaxies undergoing a collision, close passage or merger with another galaxy, or have recently been through such an episode. In contrast to the Irr 1 galaxies that are often best described as "a mess", some Irr 2 galaxies are incredibly beautiful (Fig. 1.9).

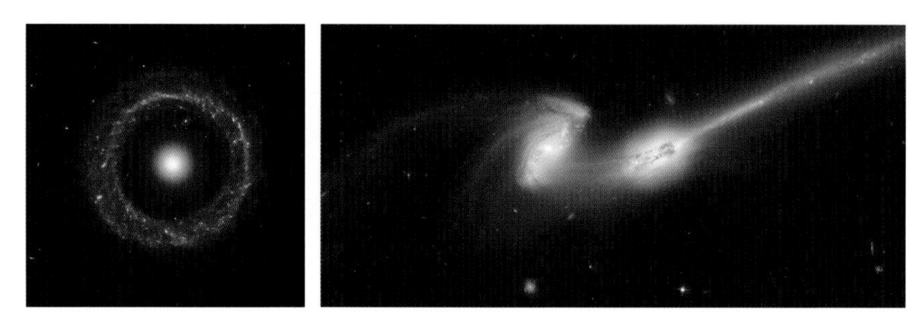

Figure 1.9 Irr 2 galaxies. Left: an HST image of Hoag's object, an example of a ring galaxy that may have resulted from a collision between a spiral galaxy and a smaller compact galaxy. (Image courtesy of Ray Lucas, NASA and the Hubble heritage team (STScI/AURA).) Right: an HST image of "The Mice" (NGC 4676), a pair of colliding galaxies that are also sometimes classed as peculiar spiral galaxies. (Image courtesy of NASA, H. Ford (JHU), G. Illingworth (UCSC/LO), M. Clampin (STScI), G. Hartig (STScI), the ACS science team and ESA.)

Peculiar galaxies are not a separate Hubble class, but an addition to his system. They are galaxies that can be classified under Hubble's scheme but also possess exceptional features not found amongst typical examples of their group. They are denoted by a "p" or "pec" being added to their Hubble classification.

Hubble's classification system has stood the test of time because of its simplicity and because the classes reflect physical differences amongst galaxies as well as their differing morphologies (Table 1.1). Thus the stars forming elliptical galaxies are mostly old, cool and have low masses (sometimes called Population II stars). On color images elliptical galaxies have a reddish hue arising from the low surface temperatures of their constituent stars. Elliptical galaxies generally contain very little (1% or less) material in the form of gas and dust, though there can be exceptions to this. The nuclei of spiral galaxies are also made up from Population II stars, but the disks contain a high proportion of gas and dust (up to 15% by mass), many young stars (Population I stars) and star-forming regions. The higher mass young stars are very hot and very luminous. Their emissions dominate the disk giving it a bluish tinge on color images. Irregular galaxies generally contain large amounts of gas and dust together with both Population I and II stars.

Table 1.1 Summary of the typical properties of galaxies.

	Elliptical	Spiral	Irregular
Size (ly)	1,000–500,000	5,000–200,000	1,000–100,000
Size (pc)	300–150,000	1,500–60,000	300–30,000
Mass ($M_\odot$)	10^5–10^{13}	10^8–10^{12}	10^7–10^{11}
Mass (kg)	10^{35}–10^{43}	10^{38}–10^{42}	10^{37}–10^{41}
Luminosity ($L_\odot$)	10^5–10^{11}	10^8–10^{11}	10^7–10^{10}
Luminosity (W)	10^{32}–10^{38}	10^{35}–10^{38}	10^{34}–10^{37}
Absolute magnitude	(-8^m)–(-23^m)	(-15^m)–(-23^m)	(-13^m)–(-20^m)

1.2.2 Galaxy Formation

The way in which the differing types of galaxies are formed is still not well understood, although clearly the basic process must be one of gravitational concentration of dispersed material. There is a striking similarity between elliptical galaxies and the nuclei of spiral galaxies that has led some astronomers to suggest an evolutionary link between the two. The gravitational field of an elliptical galaxy embedded within a dense part of the intergalactic medium would lead to material from its surroundings being attracted towards the galaxy. Conservation of angular momentum would then cause the incoming material to accumulate into a disk around the elliptical galaxy. Finally, formation of spiral arms within that disk would lead to the production of a spiral galaxy. Conversely, computer modeling of collisions and mergers between galaxies suggests that when things have stabilized, whatever might have been the types of the galaxies involved originally, the eventually resulting galaxy is an elliptical one. If these two processes do underlie galaxy formation, then since the formation of spiral galaxies will cease when most of the intergalactic matter has been consumed but galaxy collisions will continue, we must conclude that elliptical galaxies will eventually be the only ones left in the universe.

1.2.3 The Milky Way Galaxy

Legend has it that the Milky Way which we see in the sky results from milk spilled from the goddess Hera's breast when the infant Hercules was suckling. While we now know that it is actually formed from tens of thousands of stars too faint to be seen individually with the unaided eye, the legend lives on in the word "galaxy" itself since that is derived from the Greek word for milk – γαλα (gala). The Milky Way, seen from a good site, is a diffuse and irregular band of light some 10–20° across and running right the way around the sky. Galileo was able to resolve it into stars during some of his earliest telescopic observations. In the eighteenth century Sir William Herschel attempted to find the Milky Way's physical shape through a process he called "star gauging". Although he knew that stars were of differing intrinsic brightnesses, he reasoned that with enough stars, their average brightness would be constant wherever he looked in the sky. Thus if he saw more stars per unit area in one direction compared with another, then it must be that the Milky Way extended further into space in the first direction compared with the second. In this manner he found that the three-dimensional shape of the Milky Way in space was a thin rectangular box split at one end and centered on the solar system (Fig 1.10).

Herschel's star gauging is a crude but nonetheless valid method of trying to find out the Milky Way's shape. Since we know that the solar system is actually situated towards the outer edge of a spiral galaxy, not at the center of a box, why did he get such an erroneous result? The answer is because the Milky Way that we see in the sky is only a small portion of the whole Milky Way Galaxy. The distance that we can see within the region occupied by the Milky Way is limited by small

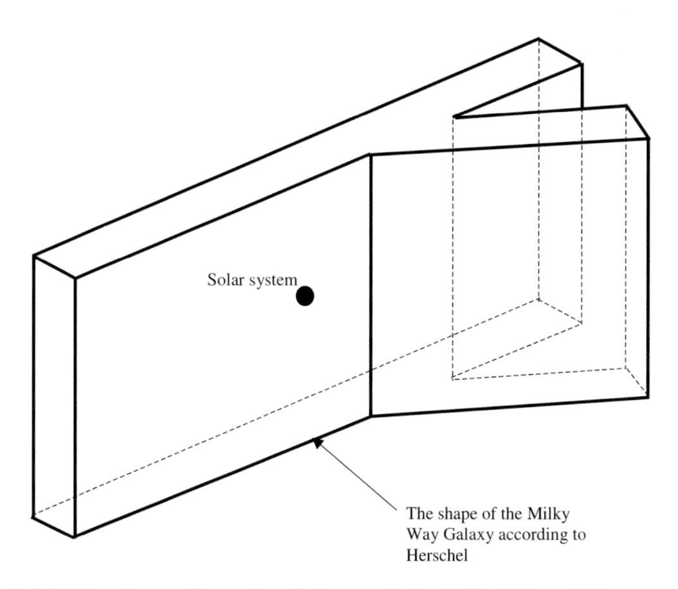

Figure 1.10 The three-dimensional shape of the Milky Way Galaxy as deduced by Sir William Herschel in the eighteenth century.

particles called interstellar dust that are composed of ices, carbon and/or silicate minerals and which absorb radiation at most wavelengths. On average along the plane of the Milky Way, the brightnesses of distant stars are dimmed by a factor of 10 for every 8,000 ly (2,500 pc) that they are away from us. However much of the dust is gathered together into huge clouds and they can cause the stars behind them to be a million or more times fainter than they would be without the cloud being present. For every 10 billion photons coming towards us from the center of the Milky Way Galaxy, for example, only one actually arrives – the other 9,999,999,999 photons are absorbed by the dust particles. Dust clouds thus hide most of the Galaxy from our sight and Herschel's "box" is in fact a small part of the spiral arm containing the solar system.

It was not until 1918 that evidence of the real size and shape of the Milky Way Galaxy began to be found. Ironically it was Harlow Shapley, who would argue for the spiral nebulae being small and local in the "Great Debate" two years later, who noticed that most globular clusters were in the southern part of the sky towards Sagittarius. He measured the distance to some of them and found their three-dimensional distribution to be a sphere centered far away in Sagittarius. He assumed that the center of the sphere must coincide with the center of the Milky Way Galaxy, and so deduced a diameter for the Galaxy of 200,000 ly (60 kpc) – about twice the modern estimate.

The true shape of the Milky Way Galaxy can only be observed directly at wavelengths that are not absorbed by the interstellar dust – principally radio and infrared radiation (Fig. 1.11). Thus little further progress could be made until radio astronomy began to develop after the Second World War, although in 1951 William Morgan found signs of spiral arms from the spatial distribution of extremely hot and bright stars. Very shortly after Morgan's work however, Dutch and Australian radio astronomers produced the first map of the Galaxy. This was obtained using the emission line from neutral hydrogen atoms at a wavelength of 0.211 m (usually referred to as the "21 centimeter line"). It actually showed the regions of greatest hydrogen gas density, but this was expected to be not too dissimilar to the spiral arms as delineated

Figure 1.11 An infrared false color image of the central regions of the Milky Way Galaxy. The image covers a region 900 ly (300 pc) across. Old and cool stars are shown in blue, while dust clouds are shown as red. The central white spot marks the center of the Galaxy where a massive black hole is hidden. (Image courtesy NASA/JPL–Caltech, S. Stolovy (SSC/Caltech).)

by the stars, so could be taken to indicate the general appearance of the Milky Way Galaxy from outside – and indeed it was a spiral galaxy.

Modern observations have refined this picture of the Galaxy although some details remain to be sorted out. The central nucleus is ellipsoidal, around 20,000 ly (6 kpc) in diameter and 6,000 ly (2 kpc) vertically. It has a 27,000 ly (8.2 kpc) long bar running through the nucleus that is aligned with its long axis towards us. The nucleus and bar are embedded in a disk containing the spiral arms that is about 100,000 ly (30 kpc) in diameter and typically 2,000 ly (600 pc) thick. The Hubble class of the Galaxy is probably SBc. The Sun is a part of the Orion spiral arm and at 25,000 ly (7.5 kpc) from the center is about half way towards the Galaxy's outer edge. The total mass of these visible parts of the Galaxy is around 100 billion solar masses of which 90% is in the form of stars and 10% is interstellar gas and dust. However the Galaxy is embedded in a halo of material that is spherical and perhaps 150,000 ly (40 kpc) in diameter. The globular clusters form part of the halo, but much of its material is in the form of dark matter and so not directly detectable (its presence is inferred from the way in which stars orbit the galaxy). The total mass of the halo is 10 times that of the visible parts of the Galaxy at around a trillion solar masses. At the center of the Galaxy and detectable as the radio source Sgr A* lies a black hole with a mass up to four million times that of the Sun. The total amount of energy emitted by the Galaxy (its luminosity) is about 1.4×10^{37} W ($3.6 \times 10^{10} L_{\odot}$) about two-thirds of which is radiated at optical wavelengths, while most of the rest is in the infrared.

1.3 The Expanding Universe

1.3.1 Hubble's Law

Whilst studying the shapes of galaxies and so devising his classification system, Hubble also set about determining their distances and radial velocities. The distances he deduced by the same method that he had used for M 31 – by measuring the periods of Cepheid variables within the galaxies. Radial velocities he obtained from the Doppler shifted wavelengths of lines in the galaxies' spectra (Boxes 1.1 and 1.2). By 1929 he had accumulated the distances and velocities for a couple of dozen galaxies. Almost all were moving away from us. The main exception was M 31, which we now know to be moving towards the Galaxy as a whole at about 50 km/s (its velocity seen from the Earth is about 300 km/s, but this includes the solar system's movement around the Galaxy). Even more remarkably, there was an overall trend to his results with the more distant galaxies having the larger redshifts. He found that for every Mpc (3.3 Mly[5]) away from us, a galaxy's redshift increased by an amount corresponding to a speed of 500 km/s (but see note in Box 1.2 about cosmological redshifts). A galaxy 3 Mpc (10 Mly) away would thus be moving away from us at about 1,500 km/s, one 5 Mpc (17 Mly) away at 2,500 km/s, and so on. The recessional velocity was thus proportional to the distance and the constant of proportionality – 500 km/s per Mpc (150 km/s per Mly) – known as Hubble's constant, has the symbol H_0. Expressed mathematically we get Hubble's law; $v = H_o D$, where v is the recessional velocity in km/s and D the distance in Mpc.

H_0 is one of the most important numbers involved in galaxy studies and cosmology. Hubble's first estimate was rapidly found to be much too high. A large amount of effort has since been expended on pinning down the value of H_0 but for many years the best estimates could only place it somewhere between 50 km/s per Mpc and 100 km/s per Mpc (15–30 km/s per Mly). In the last year or two however, results from the Wilkinson Microwave Anisotropy Probe (WMAP) have fixed its value at close to 71 km/s per Mpc (22 km/s per Mly).

Box 1.2 Doppler Shifts

When an object emitting a spectrum, such as a star or a galaxy, is in relative motion towards or away from the observer, the wavelengths of the lines (Box 1.1) in the observed spectrum change. The wavelengths become longer when the relative motion is apart, and shorter when the object and observer move towards each other (Fig. 1.12). The effect is called the Doppler shift after Christian Doppler, an Austrian physicist who studied the same effect for sound waves. The change in the wavelength depends upon the relative speed along the

[5]Light years are generally the primary unit used in this book, but the use of Mpc in this context is so universal that it will be followed here.

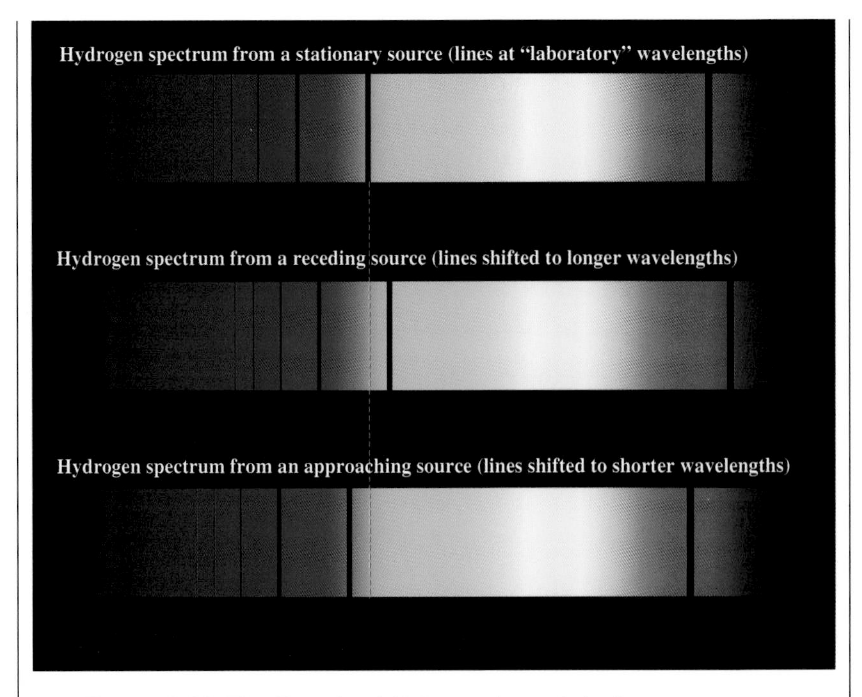

Figure 1.12 The Doppler shift for a schematic hydrogen spectrum.

line of sight (normally called the radial velocity). For speeds less than about 10% of the speed of light (i.e. < 30,000 km/s) the change in wavelength ($\Delta\lambda$) is given by the original wavelength (λ) multipliedby the relative speed (v) and divided by the speed of light (c): $\Delta\lambda = \frac{\lambda v}{c}$. At higher speeds, relativistic effects become important and the change in wavelength is larger than might be expected from the simple Doppler formula (Fig. 1.13). Radial velocities towards the observer are conventionally regarded as having negative values while velocities away from the observer are positive.

Within the Galaxy, relative speeds rarely exceed a few hundred kilometers per second, so the change in the wavelengths of spectrum lines is small – at 100 km/s, for example, the red line of hydrogen

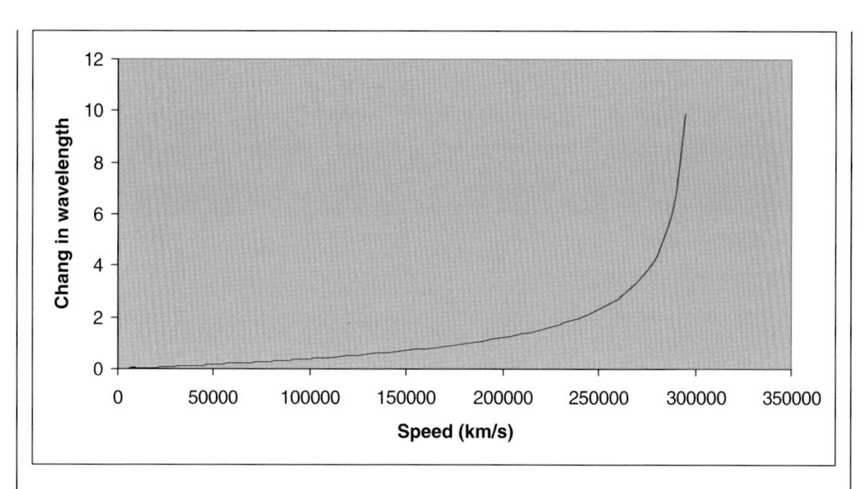

Figure 1.13 The Doppler effect at relativistic velocities. The plot shows the factor by which the wavelength changes at a particular speed. It becomes infinite at the speed of light.

moves from 656.3 nm to 656.5 nm. The wavelengths of lines in the spectra of distant galaxies are altered by much larger amounts than this, and in almost all cases the wavelengths are increased. Within the visible part of the spectrum the lines are thus moved towards the red, and so the effect is called the redshift. Of course, spectrum lines in the infrared or radio regions are moved away from red – but the effect still continues to be called the redshift. The value of the redshift is given by the change in wavelength divided by the original wavelength and is denoted by the symbol "z": $z = \frac{\Delta\lambda}{\lambda}$. For the red hydrogen line cited above we thus have $z = 0.00033$. Galaxies have now been found where the redshift exceeds a value of seven. From Fig. 1.13, such a redshift converts to a recessional speed of 290,000 km/s, 97% of the speed of light. It should be noted however that such cosmological redshifts actually arise from the expansion of the universe (see main text) and are thus caused by the expansion *of* space, not the movement of galaxies *through* space. However for the purposes of this book the distinction can be ignored and the redshifts regarded as arising simply from the speeds of the galaxies.

An important aspect of the Doppler shifts for active galaxies is the effect arising from the individual speeds of atoms or ions producing spectrum lines. Within a gas, such particles are moving around in all directions. By human standards the particles' speeds are high, 16 km/s for a hydrogen atom at 10,000 K for example. For most particles, a component of their speed will be along the line-of-sight. If that particle emits or absorbs radiation then it will do so at the Doppler shifted wavelength appropriate to its line-of-sight velocity. With large numbers of particles, some will be moving directly towards the observer and some directly away with most moving along in-between directions. Radiation will thus be absorbed or emitted over a range of wavelengths varying from the maximum redshift corresponding to the particles' speed to the maximum blueshift. The situation is complicated slightly through the particles having a variety of speeds, not just a single value, but since the average speed increases with temperature, then essentially the higher the temperature of the gas, the broader will be the resulting spectrum lines. This effect is called Doppler broadening. It should be noted that there are many other effects that can broaden spectrum lines and in particular for high gas densities, such as occur in stars, the pressure of the gas is important. With pressure broadening, the higher the gas pressure, the broader the line, until in the limit the lines are broadened completely across the spectrum and disappear, leaving the continuous spectrum (Fig. 1.3) characteristic of very dense gases, liquids and solids. Sometimes the effect broadening a spectrum line produces a particular shape (profile) for the line. With high-quality spectra it may then be possible to determine the separate contributions of the different broadening effects to the final line shape.

Doppler broadening can also arise through large-scale motions. If the individual parts of a larger object (for example small cloudlets making up a larger cloud) have intrinsic and differing motions *and* the cloud is not angularly resolved, then the observed spectrum line will be the sum of the spectrum lines from all the cloudlets, each of which will have its own speed along the line of sight and so its own Doppler shift. Where the motions of the components of the larger object are coherent or organized in some manner, such as within a rotating or expanding object, the final line profile may be of a recognizable shape (semi-elliptical in the case of a rotating star for example) and so the broadening can either be used as a measure of the rotation or expansion velocity (etc.) or the broadening effect can be removed from the line profiles so that other physical process affecting the line profile can be studied.

1.3.2 How Old Is the Universe?

Hubble's law implies that almost all galaxies are rushing away from us (the few that are approaching, like M31, are nearby and their individual motions through space overwhelm their cosmological recessions). This flight away from us however is not the result of the Milky Way Galaxy occupying a particularly undesirable part of the universe! Astronomers within any galaxy would also see all other galaxies moving away from *them*. The distances between all galaxies are thus increasing and so the universe as a whole is expanding. This mind-blowing result underlies all modern cosmological ideas and is an incredible example of the power of modern science. It has an even more mind-blowing implication though – if the universe will be bigger tomorrow than it is today, then it must have been smaller yesterday than today. Following that thought to its logical conclusion implies that at one time all the matter and other stuff within the universe must have been compressed into a very small volume (10^{-35} m across if some cosmologists have got it right) that then exploded into the Big Bang (Box 1.3).

Box 1.3 The Big Bang

The fundamental argument for the Big Bang theory of cosmology is that given in the main text – if one imagines time running backwards then the distances between the galaxies shrink and eventually they must all be crammed together into a very small volume. Although a few cosmologists still argue for other theories, the Big Bang has won most astronomers over because of three observational supports. The first is the expansion of the universe itself that arises naturally as a result of an explosive birth. The second is that the Big Bang theory predicts that the universe should be filled with enormous numbers of photons left over from its origin. Those photons were first detected in 1965 by Arno Penzias and Robert Wilson in the microwave region of the spectrum and are known as the Microwave Background Radiation (MWBR).

The third observational support for the Big Bang lies in the abundances in the universe of helium and deuterium (an isotope of hydrogen containing a proton and a neutron, also known as heavy hydrogen). By mass, hydrogen makes up about 74% and helium 24% of all the matter in the universe. Helium is produced from hydrogen during the nuclear reactions powering most stars, but in the 14 aeons or so since the Big Bang, only about 1% of the hydrogen would have been so converted, and most of that is still hidden deep inside the stars' cores. However helium is also produced in the conditions that would have existed soon after the Big Bang. The amount of helium produced does not depend very much on exactly how hot and dense those conditions were, but upon the reaction rates between sub-atomic particles. Those reaction rates can be measured, and – wonder of wonders – their values predict that around 24% of the matter in the universe should be helium! The amount of deuterium in the universe is similarly consistent with the amount predicted to be formed during the Big Bang.

There are however problems in reconciling the Big Bang to other observations such as the uniformity of the MWBR, the lack of particles known as monopoles that theoretically should be produced in huge numbers during the Big Bang and how close to Euclidean the geometry of space seems to be. These problems have led to the idea that very soon (around 10^{-35} seconds) after the Big Bang the size of the universe increased far more rapidly than its "natural" rate of expansion. This period of inflation boosted the size of the universe by perhaps as much as a factor of 10^{50}, so that the presently visible universe is only an extremely tiny fraction of the whole. More recently evidence has arisen that the visible matter and radiation present in the universe amounts to only some 5% of the whole. The remaining "stuff" making up the universe is dark matter (around 25%) and dark energy (around 70%). Dark matter could take the form of small stars, black holes, etc. that are currently undetectable or be some exotic new particle such as WIMPs (Weakly Interacting Massive Particles). As for dark energy, its nature is a complete mystery. The need to invoke such extreme processes and weird entities leaves many astronomers doubtful that we have currently "got it right" as far as the Big Bang is concerned.

If the universe has expanded at a constant rate, then it is easy to calculate from H_0, how long it is since the Big Bang occurred. For a value of H_0 of 71 km/s per Mpc, that time is 14 aeons. However it is likely that the rate of expansion of the universe has varied – slowing down through the dragging effects of gravity at least and perhaps speeding up some five to six aeons ago as suggested by recent supernova observations. Thus the "age" that we have just obtained of 14 aeons is not the actual time since the Big Bang. It is called the Hubble time and it will differ from the true age of the universe in a way that depends upon the cosmological model that is assumed. The best current estimates place the true age of the universe at a little under the Hubble time – perhaps around 13¾ aeons.

The expansion of the universe complicates the study of galaxies. As well as seeing distant galaxies as they were many aeons ago (see Box 1.4), the cosmological redshifts of their spectra mean that the radiation that we now observe was originally emitted at far shorter wavelengths. The radiation seen in the visible spectrum of a quasar at a typical redshift of three would thus have been well into the ultraviolet when it was emitted. Hence comparing nearby and distant galaxies is a problem since we are not comparing like with like.

Box 1.4 Look-Back Time

One of the wonders of astronomy for many people is the thought that they may be seeing things that no longer exist. Distances are so large that light takes an appreciable amount of time to travel to us and during that interval the object may have disappeared. Most objects of course will just change or move by small amounts, but disappearance is possible. In 1987 a hot blue star in the Large Magellanic Cloud (LMC) became a supernova – in effect exploding and disappearing. Now the LMC is some 170,000 ly (50 kpc) away from us, so the star had not actually been in existence for 170,000 years – yet it remained visible to terrestrial observers until 1987.

The interval of 170,000 years is the look-back time for the LMC. In general we see an object x million light years away from us as it was x million years ago. With occasional exceptions like the LMC supernova, look-back times within the Galaxy and for nearby galaxies are short compared with the times scales over which changes occur within galaxies. We thus observe our immediate intergalactic neighborhood more or less as it "actually" is today. This is not true for very distant galaxies – they will be quite different now compared with how we see them. The majority of quasars for example have redshift values (Box 1.2) between two and three, corresponding to distances between 11,000 Mly and 12,000 Mly (3,400–3,700 Mpc) and so to look-back times of 11–12 aeons. The black hole at the center of a bright quasar however will consume some thousands of millions of solar masses of material in an aeon, effectively emptying the central regions of its host galaxy. The properties of the quasar must therefore change significantly over a similar period.

The most distant object known at the time of writing, a gravitationally lensed galaxy, has a redshift of 7.7 and so we see it as it was 13.5 aeons ago, or only about 250 million years after the Big Bang. Surprisingly however we can look a lot further back than that. The microwave background radiation was produced at a redshift of about one thousand – just 300,000 years after the Big Bang – yet can be detected even now using relatively unsophisticated apparatus not much different from a domestic television receiver.

2
Starburst Galaxies

Summary

- The main types of active galaxy – starburst and AGN.
- The distinguishing properties of starburst galaxies.
- How starburst galaxies are detected – infrared, ultraviolet and emission line searches.
- The formation of starburst galaxies and their energy sources – buried quasars, star-forming regions, galaxy collisions and mergers.
- Starburst lifetimes.
- Boxes
  Star formation, young stars and spiral arms
  H II regions

The classification of stars
Gravitational lensing
Thermal radiation.

2.1 Recognizing Starburst Galaxies

Active galaxies take two differing forms – those whose spectra are dominated by radiation from recently formed stars, either seen directly or inferred after the stars' radiation has been absorbed and re-emitted at different wavelengths, and galaxies whose spectra are dominated by non-stellar energy sources (often called non-thermal radiation – Box 3.1). Many astronomers regard the latter type as being the "real" active galaxies and take little interest in the first type. Nonetheless the first type, that goes by the generic name of starburst galaxies, undoubtedly have more going on inside them than classical galaxies, and so are worthy of inclusion in the group, furthermore the second type of active galaxy often has starburst activity occurring as well as its non-thermal emissions. The second type of active galaxies are generally called active galactic nuclei (AGNs), because most of the activity usually occurs close to the center of the galaxies' cores. AGNs are discussed in Chap. 3, *et seq.*; here we are concerned with the starburst group.

A starburst galaxy is one that is experiencing a torrent of star formation far more intense than that found amongst the classical or normal galaxies. There is no sharp division however between classical and starburst galaxies – star-forming activity in galaxies varies more or less smoothly from galaxies with rates far less than that of the Milky Way to that of the most violent starburst galaxies.

Given that there is no sharp transition between classical and starburst galaxies, recognizing to which of the types a particular galaxy belongs is not a precise process. Several criteria can be used to refine the choice, but in the borderline region one might as well toss a coin. Firstly, starbursts usually, but not always, occur towards the center of the galaxy. Secondly, the starburst region is small compared with the size of the galaxy – typically less than 10% of its size or under 3,000 ly (1,000 pc) and with much of the activity occurring in numerous much smaller regions each a few tens of light years across (~10 pc) that have luminosities up to 100 million times that of the Sun. The energy emitted by the massive stars in these regions dominates the emission from the whole galaxy (Box 2.1) especially for the most energetic galaxies. Thirdly, the rate of star formation greatly exceeds the rate sustainable over the galaxy's lifetime, so that the starburst event must be a relatively transient episode. A classical galaxy like our own has a star-formation rate within the disk of a few solar masses per year, equivalent to perhaps 100 actual new stars since low-mass stars are far more common than those with high masses; the star-formation rate in a starburst galaxy can up to a thousand times higher than this. The starburst is usually formed from a large number of smaller star-forming regions and some of these will have progressed to the H II region stage (Box 2.2). The spectrum of a starburst galaxy thus shows the emission lines characteristic of H II regions as well as the normal absorption-line spectrum of a classical galaxy – in fact an alternative name for a starburst galaxy is H II galaxy. The emission lines in the visual region are principally due to neutral hydrogen and helium plus forbidden lines (Box 1.1) from singly ionized oxygen and nitrogen and doubly ionized oxygen and neon. Several of these lines may also be present in the spectra of the AGN type of active galaxies but the latter can be distinguished from starburst galaxies because at least some of their emission lines will be considerably broader than either the emission or absorption lines of starburst and classical galaxies.

The rate of star formation within starburst galaxies is so high that a quarter of all high-mass stars in the local universe[6] originate within

[6]This is not the same as the local cluster of galaxies that includes the Milky Way Galaxy and M 31, but is a somewhat ill-defined region over which conditions are roughly similar to those closer to the Milky Way. It probably extends out to a redshift of one, i.e. to a distance of approximately 8,000 Mly (2,500 Mpc).

them. It is likely that they also produce many of the lower mass stars, but these are too faint to detect and so that cannot be confirmed observationally. During the early stages of the universe the contribution of starbursts to the overall production of stars may have been even higher. The extreme luminosities of high-mass stars (Box 2.1) mean that about a tenth of all the energy produced within the local universe comes from starburst regions.

The energy from starbursts, particularly that coming from stellar winds and supernovae, heats the interstellar gas. The hot gas then expands outwards along the lines of least resistance, which will usually be perpendicular to the galaxy's disk. The resulting galactic winds can reach temperatures of several million kelvin, and speeds of up to a thousand kilometers per second. In many cases the wind will produce low-density bubbles up to 100,000 ly (30 kpc) across on either side of the starburst galaxy that can be detected by their x-ray emissions. It is probable that the material in galactic winds from starburst galaxies eventually merges with the intergalactic medium and is the source of the elements heavier than hydrogen and helium that are to be found there.

Starburst galaxies are located at all distances, implying that they have been formed throughout most of the life of the universe. Some, like M 82 at 11 Mly (3.3 Mpc) away from us, are nearby, and so their starburst activity is continuing at the current time. Other starburst galaxies may be found that were in existence just 1.5 aeons after the Big Bang – i.e. at distances up to 12,000 Mly (3,500 Mpc) and when perhaps one galaxy in eight contained a powerful starburst. The nearest starburst galaxy is NGC 253, which is just 8–10 Mly (3 Mpc) away from us.

Box 2.1 Star Formation, Young Stars and Spiral Arms

Anyone looking at images of spiral galaxies might well come to the conclusion that the material in the disk of the galaxy is concentrated into the arms. In fact the density of the disk is relatively uniform, there is only a slight increase in the density in the regions occupied by the arms. The reason why the arms stand out so sharply from the rest of the disk is that they contain numerous star-forming regions. In those regions the massive hot young stars are extremely bright and outshine the far more numerous less massive stars. Few star-forming regions are to be found in the remainder of the disk and so, despite there still being many stars in those regions, they appear dark in comparison with the arms.

Most stars are formed inside giant molecular clouds (GMCs). GMCs are the largest single structures found in the Galaxy, yet they have only been discovered relatively recently. Their obscurity is due to their temperatures that can be as low as 10 K (–263°C). They radiate therefore mainly in the far infrared and microwave regions of the spectrum and so their study had to await the development of suitable detectors and telescopes for those regions. GMCs can be up to 300 ly (100 pc) in size and contain between a few hundred thousand and ten million solar masses of material. About 1% of a GMC's mass is in the form of dust and so sometimes they can also be detected as dark absorbing regions. They contain, as their name suggests, a large number of different molecules many of which are organic (but produced "naturally" not as the result of the actions of living organisms). The average density of a GMC is around a billion molecules and atoms per cubic meter – for comparison our atmosphere is some 100,000 trillion times thicker than this at sea level. Within the GMC are denser regions where the numbers of particles per cubic meter can be a hundred or a thousand times the average. The Galaxy contains at least 3,000 GMCs, the nearest being in Orion and about 1,500 ly (500 pc) away from us (Fig 2.1). The visible Orion nebula (M 42, Fig 2.1) is a small part of the whole Orion GMC that has been heated to high temperatures by the hot stars inside it.

The low temperatures and relatively high concentrations of the material in the denser regions of GMCs lead to the gravitationally induced collapse of the region. Eventually that collapse produces new stars. Currently there are two models for how this may occur. In the first of these, called "gravitational collapse and fragmentation",

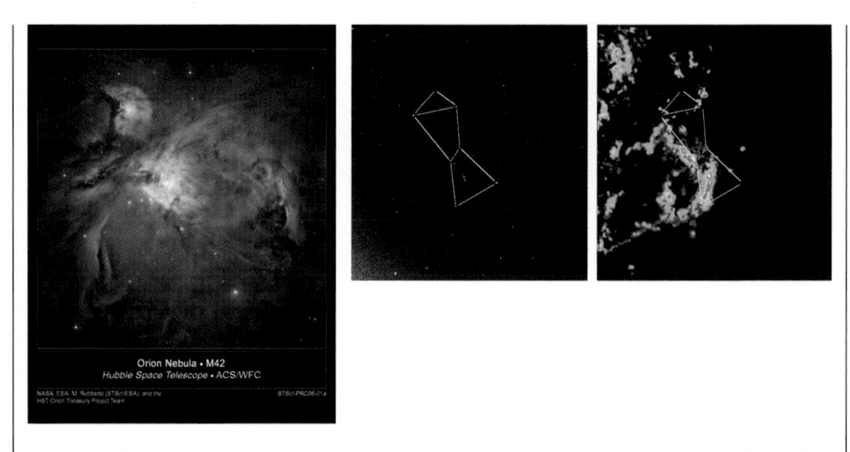

Figure 2.1 The Orion Nebula (M 42) (Left: Image courtesy of NASA, ESA, M. Robberto (STScI/ESA) and the Hubble space telescope Orion treasury project team) that is a small part of the whole Orion Giant Molecular Cloud (Right: Image courtesy of Thomas Dame, Harvard-Smithsonian Center for Astrophysics).

the collapsing clumps are large and they fragment during the infall so that many binary and multiple stars result. The second theory, "competitive accretion", has several centers, each about 1 ly (0.3 pc) across, developing within the clumps as they collapse. The centers themselves then collapse under gravity and build up their masses by accreting material from the larger surrounding clump.

Whichever of these processes is correct – and they both could be – the final result is the formation of a group of stars deep inside the remainder of the dense core of the GMC. Up to this point little of the process is observable. Now however the stars heat the dust in the surrounding material until it is hot enough to emit infrared radiation. The GMC is largely transparent at infrared wavelengths and so the young stars appear as infrared sources within the GMC. Later the energy from the stars is sufficient to heat and drive off the remainder of the GMC, and the stars become directly visible to visual observers. The hot gas before it is lost completely forms an H II region, which is a gaseous nebula with an emission-line spectrum (Box 2.2). H II regions are some of the most spectacular sights in the sky – rivalling galaxies – and the well-known Orion Nebula (M 42, Fig. 2.1) is exactly just such a region that has developed on the nearest side to us of the Orion GMC.

The young stars, that can now be seen directly, have masses ranging from 0.08 solar masses (the brown dwarf limit) to slightly over 100 solar masses (beyond which stars become unstable). By far the majority of stars though are at the low-mass end of this range – 80% have masses less than that of the Sun. The higher mass stars are thus few and far between, but their brightnesses are out of all proportion greater than those of the lower mass stars – a star with a mass half that of the Sun will have a luminosity about 3% that of the Sun, while a five solar-mass star will be 600 times brighter and a 50 solar-mass star 200,000 times brighter than the Sun. The emission from the star cluster will thus be completely dominated by the few high-mass stars within it. As well as high luminosities the large stars also have high surface temperatures. Their emitted radiation is thus predominantly in the ultraviolet and blue regions of the spectrum. The ultraviolet radiation ionizes the surrounding gas, producing the H II region, while the blue emission produces the bluish tinge detectable for the spiral arms on color images.

The enormous luminosity of the young massive stars means that they consume the hydrogen available as fuel for nuclear reactions much more rapidly than the more frugal lower mass stars. A 50-solar mass star will run out of hydrogen in around two and a half million years, compared with 10 billion years for the Sun. The stars in the Orion nebula are about one million years old. The hottest of the stars forming the Trapezium at the center of the nebula has a mass of about 60 solar masses and a luminosity some 300,000 times that of the Sun and it is thus now about halfway through its life. Soon (in astrophysical terms) after the birth of a group of stars therefore, the high-mass members will come to the ends of their lives and fade away or explode as supernovae and disappear. Since those massive stars have dominated the emission from the star-forming region, as their luminosities diminish, so also will that of the star-forming region.

The spiral arms of galaxies are thus seen as such because the formation of stars is triggered at their leading edges, perhaps by a density wave or some other process. After a few million or tens of millions of years the massive stars die and the star-forming regions disappear from our sight, so that the spiral arm seems to fade away at its trailing edge. The majority of the stars produced within the star-forming-region – those with the lower masses – are still however in existence and continuing to radiate, but are too faint to be detectable across cosmological distances. The general density within the disk of the galaxy is thus reasonably uniform with only a slight enhancement near the visible spiral arms.

Box 2.2 H II Regions

Stars of spectral classes O and B (Box 2.3) have very high surface temperatures and emit a great deal of ultraviolet radiation. The lives of such stars are short (Box 2.1) and so they are usually still to be seen in the groups where they were produced by the star-forming region, along with many cooler, lower mass stars. The ultraviolet photons from the few hot stars in such a group have enough energy to ionize many of the atoms in the gas surrounding it. A bubble of very hot (10,000 K) gas, which is almost completely ionized in its inner regions, thus surrounds the group of stars. The bubble is called an H II region because the hydrogen within it is almost completely ionized, and the symbol for ionized hydrogen is H II (Box 1.1). The Orion nebula (M42, Fig. 2.1) is a relatively nearby H II region.

Within the hot bubble, the ions and electrons will be moving around rapidly and will undergo frequent collisions. During such collisions it is possible for the electron to bond with the ion (known as recombination) to produce the neutral atom again (or a lower stage of ionization). The electron's energy will be released in the form of photons belonging to some of the characteristic spectrum lines of that atom or ion. H II regions thus have emission-line spectra (Fig 1.3) and the strongest lines in the visual spectrum are usually due to hydrogen (H I 388.9, H I 397.0, H I 410.1, H I 434.0, H I 486.1, H I 656.3), helium (He I 388.8, He I 447.1), nitrogen ([N II] 575.5, [N II] 654.8, [N II] 658.3), oxygen ([O II] 372.7, [O III] 495.1, [O III] 500.7) and neon ([Ne III] 386.9).

When a starburst region is directly visible, it is the O and B stars plus the H II regions that provide the bulk of the emitted radiation. Similarly it is principally these stars and nebulae that cause the spiral arms of classical galaxies to stand out within the galaxies' disks (Chap. 1).

Box 2.3 The Classification of Stars

A useful and widely used classification system for stars is based upon their surface temperatures and is known as a star's spectral type. For historical reasons the labeling of the spectral type is illogical and inconvenient, however it seems unlikely to be changed now, so the reader of necessity has to come to terms with it. Refinements and additions have been added in recent years, but for the purposes of this book, the basic system will suffice.

The classification is based upon the stars' surface temperatures, but these are not measured directly. Instead the varying appearances of the visual spectra of stars with differing surface temperatures are used and the class determined from the presence or absence of certain spectrum lines or from the changing relative intensities between pairs of spectrum lines (hence "spectral type"). There are seven major

Table 2.1 Stellar spectral types and surface temperatures.[7]

Spectral type	Surface temperature (K)	Surface temperature (°C)	Surface temperature (°F)
O5	42,000	42,000	76,000
B0	30,000	30,000	54,000
B5	15,200	15,000	27,000
A0	9,790	9,520	17,200
A5	8,180	7,910	14,300
F0	7,300	7,030	12,700
F5	6,650	6,380	11,500
G0	5,940	5,670	10,200
G5	5,560	5,290	9,550
K0	5,150	4,880	8,820
K5	4,410	4,140	7,480
M0	3,840	3,570	6,460
M5	3,170	2,900	5,250

classes indicated by upper case letters, which in order of decreasing temperature are:

O B A F G K M

As previously mentioned, the inconvenient labels arose historically. The reader may however find the following mnemonic useful:

Oh Be A Fine Girl/Guy Kiss Me

Each of these major classes is sub-divided into 10, with each sub-division labeled by a number from 0 to 9. The Sun thus has a spectral class of G2; Sirius (α CMa) is A1 while Antares (α Sco) is M2. The hottest stars known at the moment are O3, while at the cool end, the sequence can be continued beyond M9 into the brown dwarfs. The relationship between surface temperature and spectral class is given in Table 2.1.

[7]These are the values for solar-type stars, usually called main-sequence stars. Slightly different temperatures are found for larger stars (giants and supergiants) of the same spectral class.

Additionally, the brightness of a star can often be estimated from its spectrum and so a luminosity classification is frequently added to the spectral class. The brighter stars with a given surface temperature (i.e. spectral type) are clearly larger than fainter stars of the same temperature – their emissions per unit area are the same, so the brighter stars must have larger surface areas than the fainter ones. The luminosity classification is thus also a size classification. A Roman numeral from I to VII is used for the luminosity class (Table 2.2):

Table 2.2 Stellar luminosity classes.

Luminosity class	Luminosity for spectral type G5 ($L_\odot$)	Star type
I	~23,000	Supergiant
II	~1,000	Bright giant
III	~33	Giant
IV	~5	Subgiant
V	~0.7	Dwarf (also called main-sequence stars – these are the commonest stars by far, and the class to which the Sun belongs).
VI	~0.2	Subdwarf
VII	~0.0001[8]	White dwarf.

The luminosity class is added to the spectral type so that the Sun becomes G2 V, Sirius, A1 V while Antares is M2 I.

[8]White dwarfs have a separate classification system from the normal spectral type. This is the luminosity for a white dwarf with a surface temperature similar to that of a G5 star. Most white dwarfs however have much higher temperatures than this.

2.2 Finding Starburst Galaxies

H II regions and hot O and B stars can be seen within the images of many galaxies by direct visual inspection, and galaxies with numerous such indicators of star-forming regions have long been identified – the Mice (Fig. 1.9) and M 82 (Fig. 2.2) for example. Until the launch of the Infrared Astronomy Satellite (IRAS) in 1983 however, the number of galaxies known to have such unusually vigorous rates of star formation was small. IRAS' mission was to survey the sky at infrared wavelengths and during its 10 months of operation – its lifetime was limited by the amount of liquid helium that it could carry to cool its telescope and detectors – it observed over 20,000 galaxies as well as more than 200,000 stars and other objects. The most outstanding discovery that came from IRAS' data however was of a new class of galaxies that

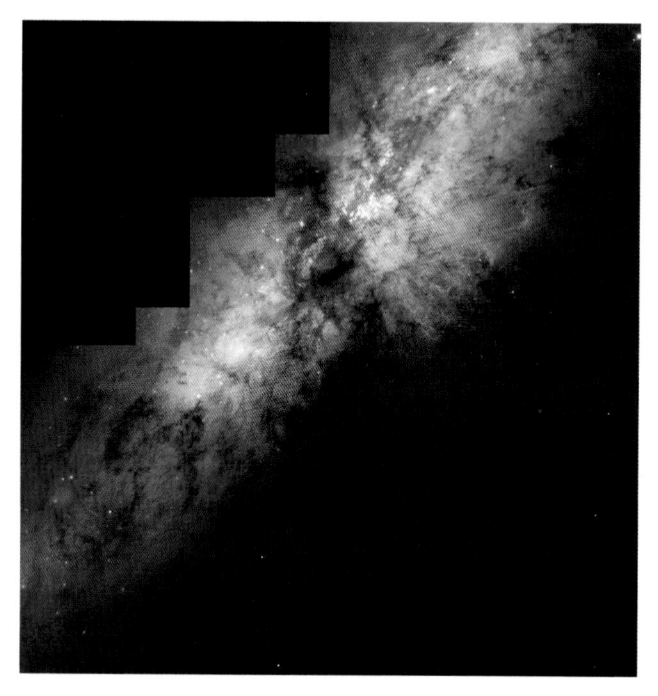

Figure 2.2 An HST image of the starburst galaxy, M 82. (Image courtesy of NASA, ESA, R. de Grijs (Institute of Astronomy, Cambridge, UK).)

emitted huge amounts of energy in the far infrared. The infrared energy emitted by these galaxies may originate from star-forming regions and in many instances greatly exceeds the total energy coming from normal large galaxies. These galaxies have become known by several names – IRAS galaxies, H II region galaxies, Far Infrared Galaxies (FIRGs), super starburst galaxies and Ultraluminous Infrared Galaxies (ULIRGs) as well as starburst galaxies. More recently SCUBA galaxies (for the Sub-millimeter Common User Bolometer Array that is used on the James Clerk Maxwell telescope) and sub-millimeter galaxies have been added to the pantheon for those galaxies whose energy is mainly found at even longer wavelengths than the far infrared. Some of these names may be synonyms, but it seems likely that more than just rapid star formation is needed to explain all the phenomena (Sect. 2.3, Chap. 3, *et seq.*).

Other than through infrared surveys – and IRAS has been followed by later infrared spacecraft such as ISO (Infrared Space Observatory) and Spitzer – starburst galaxies are mainly discovered via two of their other properties. The massive stars in star-forming regions emit much of their energy at ultraviolet wavelengths, so starburst galaxies may be identified from their high ultraviolet emissions. The ultraviolet photons also ionize atoms and the subsequent recombination of ions and electrons produces emission lines at visual wavelengths (Box 2.3). In this way in the 1960s and 1970s, surveys conducted by Benjamin Markarian using the 1.3-m Schmidt camera at the Byurakan observatory and looking for galaxies with near ultraviolet excesses and/or emission lines found over 1,500 examples. Out of these Markarian galaxies around 90% proved to be starburst galaxies while the remainder were mostly Seyfert galaxies (Chap. 3).

The brightest starburst galaxies are the ULIRGs and these may be the brightest objects in the universe (they may though not be true starburst galaxies; see Sect. 2.3). ULIRGs all have far infrared brightnesses that exceed a trillion solar luminosities. For comparison, quasars and QSOs (Chap. 3, *et seq.*) can have total brightnesses up to ten trillion solar luminosities. At least one ULIRG, however, may out-do even the quasars. FSC 10214+4724[9], an IRAS galaxy that is also known as the Rowan-Robinson galaxy after its discoverer Michael Rowan-Robinson, is some 11,500 Mly (3,500 Mpc) away from us. Its apparent far infrared luminosity is an incredible 300 trillion times that of the Sun – perhaps 30 times brighter than the brightest quasar. There is some evidence though that the apparent brightness of FSC 10214+4724 may have been enhanced, perhaps by as much as a factor of 10, by gravitational lensing (Box 2.4), so that its true luminosity is much less than $3 \times 10^{14}\,L_\odot$. It must remain, however, a strong candidate for being the brightest object in the visible universe.

Box 2.4 Gravitational Lensing

Sir Arthur Eddington undertook one of the earliest observational tests of Albert Einstein's theory of general relativity in 1919. Einstein had predicted that gravitational fields would deflect the paths of light beams. Light (or other e-m radiation) skimming the surface of the Sun should thus have its direction of travel changed slightly by the Sun's gravitational field and the object from whence the light originated should appear to be in a slightly different part of the sky. Eddington observed the solar eclipse of 1919 to see if the positions of stars then behind the Sun were moved compared with their positions when the Sun was not in that region. He found that they were moved and by just the amount required by general relativity.

[9]FSC stands for the IRAS Faint Source Catalogue. The numbers give the object's right ascension and declination: RA = 10 h 21.4 m, Dec = +47°24′. With the huge numbers of stars, nebulae, galaxies, etc. that the HST and other space observatories have identified plus contributions from ground-based surveys like the Sloan Digital Sky Survey (SDSS), using names such as "Whirlpool" galaxy or catalogue numbers such as "NGC 1275" has become impractical. This type of positional designation is thus now very widely used for objects of many types.

A gravitational field can thus deflect the paths of light beams passing through it. Now deflecting the paths of light beams is exactly what a lens does. However for a converging lens, the amount of the deflection increases the further away the light beam passes from the center of the lens (Fig. 2.3). For the Sun, or any other object where the light passes by beyond its outer limits, the gravitational field weakens away from the object and so the gravitational deflection decreases outwards (Fig. 2.3). The gravitational "lens" thus does not

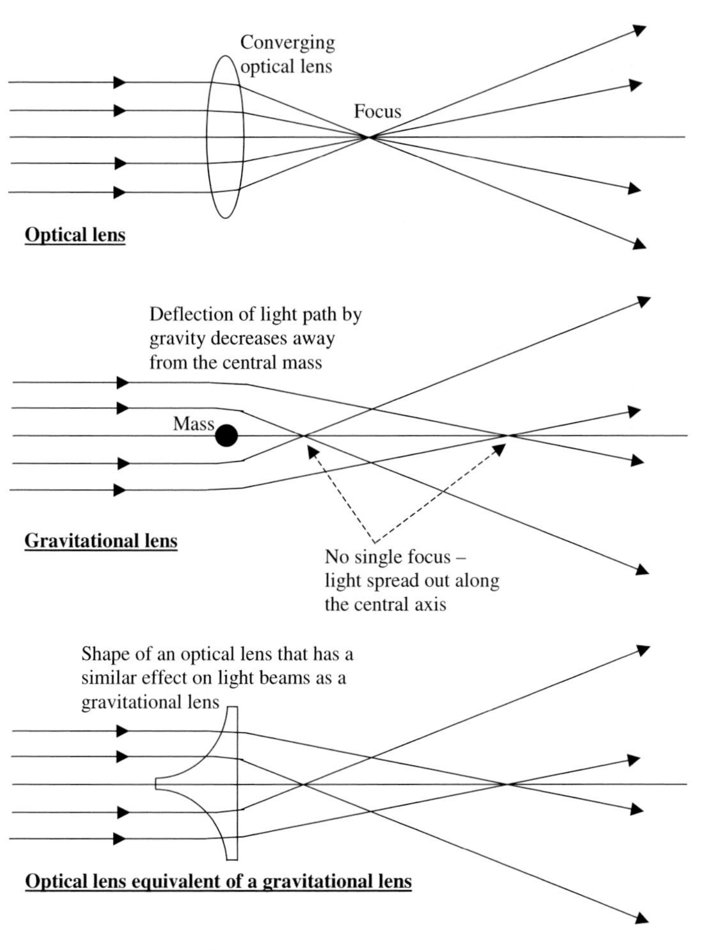

Figure 2.3 Optical and gravitational lenses.

produce an image in the manner of a converging optical lens; instead its optical equivalent would be like the stem and base of a wine glass (Fig. 2.3). In both cases no image is produced, but light from the distant object is concentrated along the central axis. To an observer on or close to the central axis the distant object would thus appear brighter than if the lensing mass were not present.

A gravitational lens does not produce a true image, but if the distant object, the lens and the observer are almost exactly aligned, then the light from the distant object will be spread out into a ring centered on the lensing object (usually a galaxy). Such rings are known as Einstein rings, and although the chances of the observer being in the right position are very small, several such rings have been found.

Much more frequently the observer is close to but not on the central axis of the gravitational lens. The Einstein ring then breaks up into one or more arcs or more distorted shapes or into several point sources. In the latter case it is possible to get four point images at the corners of a square and the effect is then known as an Einstein cross.

2.3 The Origins of Starburst Galaxies

The far infrared emissions from FIRGs and ULIRGs, etc. originate as radiation from hot dust particles (thermal emission; see Box 2.5). In most cases the starburst regions are not seen directly, but their existence is inferred as the sources of the energy that heats the dust. Since the energy source is not seen directly, it is possible that for some of these galaxies it is not a starburst region. In particular some ULIRGs have broader emission lines than most other starburst and normal galaxies – and broad emission lines are a characteristic of the AGN type of active galaxy (Chap. 3, *et seq.*). An alternative model for some ULIRGs is thus that of a buried quasar. The quasar is deeply hidden inside a dense region of molecular gas and dust whose mass may amount to ten billion times that of the Sun. The quasar gives no indication of its presence except in the one or two galaxies where a small amount of its radiation leaks out. The energy from the quasar goes to heat the dust, which then re-radiates at far infrared wavelengths just as though a starburst region had heated it. To add to the confusion however, it is quite possible that ULIRGs containing buried quasars also contain starburst regions. Recent results from the ISO spacecraft suggest that about a quarter of the observed ULIRGS are powered by hidden quasars while infrared spectroscopic observations from the European Southern Observatory's (ESO) Very Large Telescope (VLT) imply that two-thirds of ULIRGs contain AGNs although only in a third of these does the AGN contribute significantly to the overall luminosity.

Box 2.5 Thermal Radiation

The spectrum of a hot solid, liquid or dense gas is a continuous one (Box 1.1) but the emission does not occur equally at all wavelengths. There is a wavelength at which the emitted energy peaks and the intensity then drops off rapidly towards shorter wavelengths, and somewhat more slowly towards longer wavelengths. The wavelength of the peak intensity becomes shorter as the temperature of the emitting body rises. This is a matter of common experience in that a heated object initially glows a deep red, then becomes yellow and eventually white as it gets hotter. The wavelength of the peak emission for a wider temperature range is given in Table 2.3.

GMCs, whose temperatures are less than 100 K, thus emit energy at far infrared/sub-millimeter wavelengths, the dust involved in infrared galaxies has temperatures ranging from less than one hundred to a few hundred K so that their peak emissions occur in

Table 2.3 Thermal radiation peak emission wavelengths.

Temperature (K)	Wavelength of the peak emission intensity	Part of the spectrum	Example
10	300 μm	Far infrared/sub-millimeter	MWBR (2.7 K)
20	150 μm	Far infrared	
50	60 μm	Far infrared	
100	30 μm	Mid-infrared	Dust in GMCs
200	15 μm	Mid-infrared	Earth (300 K)
500	6 μm	Near infrared	
1,000	3 μm	Near infrared	Heated interstellar dust
2,000	1.5 μm	Near infrared	
5,000	600 nm	Visual	Sun
10,000	300 nm	Near ultraviolet	
20,000	150 nm	Ultraviolet	
50,000	60 nm	Ultraviolet	O-type star
100,000	30 nm	Extreme ultraviolet	
200,000	15 nm	Extreme ultraviolet	
500,000	6 nm	Soft x-ray	
1,000,000	3 nm	Soft x-ray	Solar corona
2.000,000	1.5 nm	X-ray	
5,000,000	600 pm	X-ray	Starburst winds
10,000,000	300 pm	X-ray	

the far to near infrared, the Sun at almost 6,000 K emits primarily in the visual region, while the galactic winds from starburst galaxies reach temperatures of millions of degrees and so are detectable at x-ray wavelengths.

With most, perhaps three-quarters, of the starburst galaxies, however, it is vigorous star formation that leads to their unusual properties and behaviors. Almost all starburst galaxies – including even those ULIRGs that might contain buried quasars – show signs of undergoing or of having recently undergone an encounter with another galaxy. In clusters of galaxies the separations between individual galaxies even today are typically only around 10–100 times the sizes of the galaxies. In the past, since the universe is expanding, the relative separations would have been much smaller. A large cluster of galaxies is thus in much the same situation as a thousand jumbo jets that have all to fly at the same time and remain within the same cubic mile of the atmosphere. Collisions would be inevitable – the more so because within the cluster, gravity is pulling the individual galaxies towards each other the whole time.

A collision between two galaxies though is quite different from a collision between two aircraft. The stars within the galaxies are separated by enormous distances compared with their diameters, so that they simply pass each other by. A few stars may have encounters sufficiently close to change their trajectories by significant amounts, but it is unlikely that any two stars will actually hit each other. The gravitational fields of the galaxies as a whole though will alter the paths of most of the stars in both galaxies. These tidal forces can result in the formation of many strange structures such as double nuclei; the long tails seen in the Mice (Fig 1.9) and even produce ring galaxies such as Hoag's object (Fig. 1.9).

The situation is different however, for the gas and dust clouds within the galaxies, these will actually hit each other. The outcome of the collision will then depend upon the details of the encounter – one galaxy may be stripped of its gas and dust leaving the other enriched, or the gas and dust clouds in both galaxies may be made increasingly turbulent and/or driven towards the centers of the galaxies, or sufficient energy may be transferred out to the galactic haloes for the two galaxies to merge into a single larger galaxy. Whatever the details of the interaction between the galaxies, the interstellar material within them is disturbed from its previous relatively quiescent state. Inevitably in some places that disturbance will result in an increase in density and that in turn is likely to be sufficient to trigger star formation. The more interstellar material contained within the galaxies and the more that it is concentrated into a small volume by the interaction, the stronger and more violent will be the resulting outburst of star formation. Since spiral galaxies contain the most interstellar material, interactions involving two large spiral galaxies produce the most violent starbursts and in some cases probably lead to the formation of ULIRGs.

The timescale for collisions between galaxies is typically about an aeon. The interstellar material is thus likely to continue to be disturbed for that period of time. The starburst may therefore also persist for a similar period. It seems likely however that the lifetime of the starburst will be significantly shorter than an aeon since the interstellar material will rapidly be consumed and so little will be left to fuel starbursts during the later stages of the interaction. The lifetime of a typical starburst is thus probably limited to a few hundred million years.

Starbursts can be found within galaxies that do not appear to have been through a recent encounter with another galaxy. In these cases, the galaxies frequently possess prominent central bars. The tidal effect of the bar can draw interstellar material in towards the center of the galaxy, increasing its density and so triggering a starburst. There remain however other starburst galaxies, especially amongst the weaker starbursts occurring in small galaxies, where the cause of the starburst is still a mystery.

3

Active Galaxies

Summary

- AGNs radiate predominantly by non-thermal processes.
- Basic properties of AGNs in general.
- Preview of the unified model for AGNs – super-massive black holes and accretion disks.
- Types of AGNs and their inter-relationships.
- Review of the properties and behaviors of the principal types of AGNs:
  Seyfert galaxies
  Low-Ionization Nuclear Emission-line Region galaxies
  Quasi-Stellar Objects and quasars

Blazars
Double-lobed radio galaxies.

- Boxes

Non-thermal radiation (line emission and masers, free–free radiation, synchrotron radiation, inverse Compton scattering)
Galaxy searches and surveys
Polarization
Light size
Reverberation mapping
Error boxes and identifications
Halton Arp and non-cosmological redshifts.

3.1 What Is an Active Galaxy?

Active galaxies that are not primarily starburst galaxies (Chap. 2) are, in the opinion of many astronomers, the "real" active galaxies. The justification for this view is that in almost all cases the unusual activity may ultimately be traced back to interactions with super-massive black holes (Sect. 3.2.1 and Chap. 5) embedded within the galaxies. However it is becoming more and more apparent that most, perhaps all, galaxies contain massive black holes, even those which appear absolutely normal – our own galaxy for example has a central black hole with a mass estimated to lie between 2.6 and 4 megasuns ($MM_{\odot}$[10]), while the Andromeda galaxy (M 31) has a central black hole of about 140 megasuns. The possession of a massive black hole is thus not necessarily the distinguishing factor for this type of active galaxy. Instead the difference between a normal and an active galaxy appears to lie in the availability

[10]A million solar masses = 2×10^{36} kg – this is a term that has been invented for use in this book, which seems to the author to be a convenient and useful, though non-SI, unit. It replaces the frequently used but awkward phrases such as "a million times the mass of the Sun" that abound within discussions of AGNs.

of material (stars, planets, gas and dust – otherwise known as fuel) to power the activity by interacting with and being consumed by the black hole. More detailed discussions of these processes will be left, though, for later chapters.

As we shall see shortly there are a bewildering variety of objects considered to be non-starburst active galaxies, which have a correspondingly enormous range of differing properties. The one common and defining factor however is that a significant amount, often most, of the energy radiated by the galaxy originates from non-thermal processes (Boxes 2.5 and 3.1). To put this another way – the energy from an active galaxy does *not* primarily come from stars. However at this point we need to become more precise in our statements because with most non-starburst active galaxies, the activity is confined to a small region around and within the central nucleus of the galaxy, while the remainder of the galaxy is relatively normal. The outer parts of an active galaxy are usually thus radiating just as a result of the stars within them, it is only from the core of the galaxy's nucleus that non-stellar radiation dominates. Hence it is the almost universal practice to refer to "Active Galactic Nuclei" (AGNs) rather than to the less exact "active galaxies" for the non-starburst variety of these objects and we shall adopt this practice from now on.

Box 3.1 Non-Thermal Radiation

We have seen (Box 2.5) that any object whose temperature is above absolute zero emits radiation and that the wavelength of the peak of that emission shortens as the temperature of the emitting object rises. This is thermal radiation, familiar to everyone from tungsten filament light bulbs and many other sources. Most people however will also be familiar with fluorescent lights or yellow sodium street-lights and the radiation from these does not arise (primarily) from hot objects. The emissions from fluorescent and sodium lights are thus examples of non-thermal radiation. For galaxies the main types of non-thermal radiation that are found are called line emission, synchrotron radiation, free–free radiation, inverse Compton scattering and maser emission.

Line Emission and Masers

Strong emission lines (Box 1.1) dominate the optical spectra of AGNs – indeed it is one of their defining characteristics. The allowed energies for electrons within atoms and ions can be of only certain fixed values (for further information on this see more detailed sources on spectroscopy or quantum mechanics – see Appendix 1). If an electron whose energy is at one allowed level changes that energy, then it must end up with another allowed amount of energy. The difference between the two energies is emitted as a photon if the second energy is lower than the first, or has to be supplied from somewhere else in the reverse case. The values of the allowed electron energies are fixed within a given type of atom or ion, so the energy difference between any two levels is also fixed. Since a photon's energy is related to its wavelength (Fig. 1.2), whenever an electron within a particular type of atom or ion moves from a higher allowed energy to a lower allowed energy the photon that is emitted has a fixed wavelength corresponding to the energy difference between the two levels. When large numbers of atoms or ions have electrons simultaneously moving between the same pair of allowed energies, numerous photons with identical wavelengths will be produced and a spectrum will show an emission line at that wavelength (absorption lines arise as photons of that same wavelength are absorbed by atoms or ions to provide the energy for their electrons to move from the lower allowed energy to the higher). Most atoms and ions have many permitted energies for their electrons and the unique patterns of spectrum lines that characterize each element and its ions (Box 1.1) arise from the energy differences between pairs of these energies.

Normally atoms and ions are found in their ground states, i.e. their electrons are all in the lowest of the permitted energy levels. If emission lines are to be produced then one or more electrons must be pushed up to higher levels – a process called excitation (or ionization if the electron becomes completely separated from the atom). Energy has to be supplied to the atom or ion if it is to be excited or ionized and in astrophysical situations the two main sources for this energy are the absorption of high-energy photons or collisions with other atoms, ions or electrons. Normally, once an electron has been excited it will return to the ground state in about 10 nanoseconds. Although it can do this by going directly to the lowest permitted level, more often the electron will make several successive smaller energy changes, cascading through other excited energies on its way down. A photon will be emitted for each pair of energy levels through which the electron passes. A hydrogen atom that has become ionized through the absorption of an ultraviolet photon whose wavelength was (say) 85 nm, might thus recombine with an electron and emit photons at wavelengths of 813 nm, 4,051 nm, 1,875 nm, 656 nm and 121.5 nm. The last four emissions are at fixed wavelengths and so occur as emission lines. The first photon emitted (at 813 nm) however will form a part of an emission band – a region of emission that covers a much greater range of wavelengths than a single spectrum line. The reason for this is that a free electron can have any energy with respect to an ion. Upon recombination therefore the first energy change is from the free electron's unfixed energy to some fixed energy within the ion. The energy that is released and so the wavelength of the resulting photon thus varies from one ion–electron recombination pair to another.

Some of the emission lines in AGNs are allowed, others are forbidden (Box 1.1). Both types of lines are produced by the mechanism that has just been discussed. The difference arises because for one of the pairs of permitted electron energies in the cascade that results in the forbidden line, the electron takes much longer than the usual 10 ns to change its energy. On average, the electron might linger in the higher energy state for a few milliseconds if the emission is semiforbidden or for as long as ten million years for the most strongly forbidden lines. When an electron gets stuck in this manner, the higher permitted energy is referred to as a metastable level. Within stars and other "normal" situations, an atom or ion with an electron in a metastable level will be disturbed through absorbing another photon or undergoing a collision, etc. long before the electron changes its energy naturally, so no photon is produced and the forbidden line is very weak or non-existent compared with allowed spectrum lines. In interstellar gas clouds though, the density is very low so when an atom or ion in a metastable level is also remote from any stars it is unlikely to be disturbed by either collisions or photon absorptions. The electron will then have sufficient time to change its energy to the lower value in a conventional manner thus producing a forbidden emission line that has a normal strength.

The super-strength forbidden emission lines found in the spectra of many nebulae and AGNs arise in the way outlined above but with one additional process in operation; the atoms and ions are still occasionally being excited or ionized and also the metastable level is relatively long-lived. In this situation, whenever the electron in an excited atom or ion cascades down the route via the metastable level, it sticks at that level. After a while, the majority of electrons or ions will become excited to the metastable level and few will be left in the ground state – a situation called population inversion. Almost the only escape for the electrons in the metastable level now is via the production of the forbidden line, and so that line becomes as strong or even stronger than the most intense of the allowed emission lines.

Finally we may turn to masers, which are the microwave equivalent of lasers, the names deriving from Microwave/Light Amplification by Stimulated Emission of Radiation. Naturally occurring masers are found associated with many GMCs and AGNs. The maser emission arises from a population inversion associated with a metastable level combined with a phenomenon known as stimulated emission. In stimulated emission a photon whose energy exactly equals the energy difference between the metastable level and the lower permitted electron energy forces (stimulates) an electron in the metastable level to move down to the lower one. In the process a second photon is emitted that is exactly in step (phase) with the stimulating photon and moving in the same direction. These two photons can now stimulate two more emissions and so on until the intense, highly collimated, beam of a maser is produced. The efficiency of the stimulated emission process reduces as the photon energy increases, so the conditions for naturally occurring lasers are much less likely to occur than those for masers – and indeed natural lasers have yet to be found.

Free–Free Radiation

The basic process underlying emission of e-m radiation is that of the acceleration of an electric charge – and that includes changing the direction of motion as well as speeding it up or slowing it down. Theoretically the nature of the electric charge that is involved is immaterial but since the mass of a proton is 2,000 times that of an electron, a given force will accelerate an electron 2,000 times more strongly than a proton. In practice therefore it is almost always electrons that are the accelerated charges involved in producing radiation.

Thus if the path of a moving free electron is altered it will emit radiation – the loss of energy that this represents will also cause it to slow down, but that will not lead to further emissions. With free–free radiation, the electric field of an ion diverts the electron's path, but not sufficiently to allow the ion and electron to recombine (Fig. 3.1). The name of this radiation derives from the electron being free of the ion before the encounter and still free from it afterwards. The emissions discussed in the previous sub-section may similarly be called free–bound radiation when a free electron recombines with (i.e. becomes bound to) an ion, and bound–bound radiation when the electron is within the atom or ion both before and after it changes energy. Since the electron is slowed down by the emission

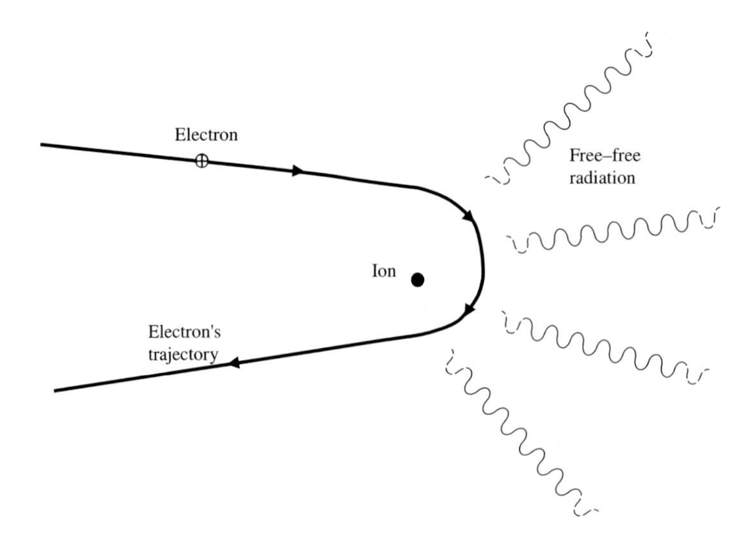

Figure 3.1 Schematic illustration of the production of free–free radiation when an electron interacts with an ion.

of free–free radiation, it is also sometimes called *bremsstrahlung* radiation after the German word for "braking".

Until densities such as those experienced inside white dwarfs are reached, the energies of free electrons are unrestricted. For free–free radiation the energy can also change by any amount short of becoming low enough for the electron and ion recombine. The radiation arising from free–free interactions between electrons and ions can thus be of any wavelength, and the free–free radiation spectrum is a continuous one. The spectrum is also flat[11], i.e. the energy emitted at any given frequency is almost constant providing that the emitting material has a density low enough to be transparent to the radiation (Fig. 3.2). At low frequencies and/or for high densities the material becomes opaque and the free–free radiation spectrum then reverts to that of thermal radiation (Box 2.5). For the temperatures normally

[11]Actually there is a very slow reduction in the intensity as the frequency increases, but only by 20% for a 10-fold increase in frequency, so regarding the spectrum as flat is good enough for many purposes.

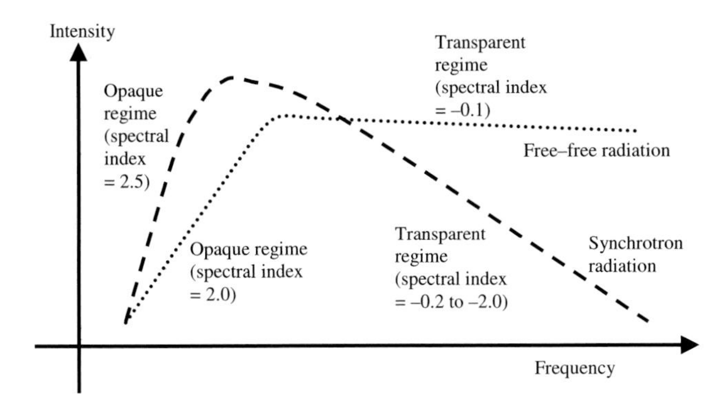

Figure 3.2 Schematic spectral energy distributions for free–free and synchrotron radiation sources.

encountered within stellar atmospheres and hot interstellar nebulae free–free emission mostly occurs at radio wavelengths, although it can extend as far as the optical region.

Synchrotron Radiation

Synchrotron radiation is one of the most important sources of radiation from AGNs. It derives it name from the particle accelerators used by atomic physicists where it can lead to serious energy losses from the particles. It is similar to free–free radiation in that it results from the paths of free electrons being altered, but this time the accelerating force arises from the electrons' interactions with a macroscopic magnetic field not with the electric field of an ion. From school physics we know that a magnetic field exerts a force on a conductor carrying an electric current within it. A moving free electron constitutes an electric current and unless it is moving directly along the magnetic field lines, the electron will experience the induced force directly. The free electron's path will thus become a spiral in the presence of a magnetic field (Fig. 3.3) and the acceleration will lead to the emission of radiation.

Although synchrotron radiation is emitted whatever the electron's speed (at slow speeds it may sometimes be called

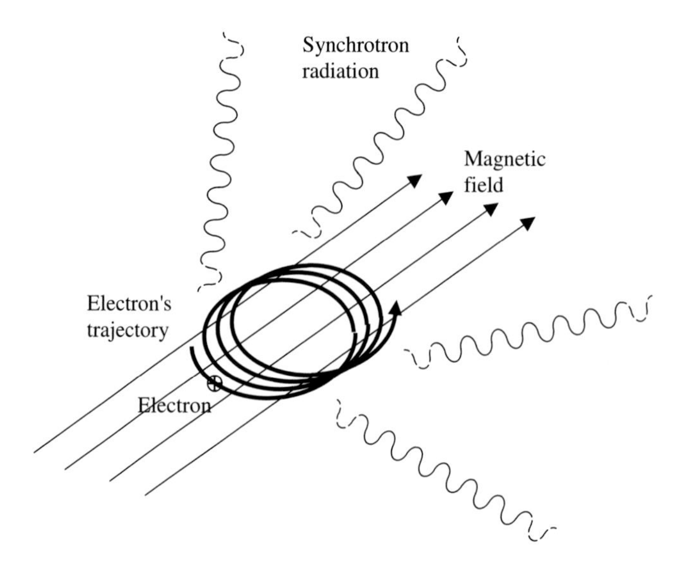

Figure 3.3 A free electron's trajectory in the presence of a magnetic field with the emission of synchrotron radiation.

gyrosynchrotron radiation), in most astrophysical situations the electrons are moving at speeds close to that of light. The radiation emitted by a single relativistic electron spans a range of frequencies and peaks at a high multiple of the frequency with which the electron is spiralling around within the magnetic field. The frequency of that peak emission depends upon the component of the electron's speed perpendicular to the magnetic field – it is about 20 times the spiralling frequency for electrons moving at $0.99c$, for example. The spectrum of a real synchrotron radiation source however is made up from the divergent individual spectra of electrons that have various speeds and whose trajectories have various angles to the magnetic field. Additionally, the magnetic field may itself alter in strength and direction over the emission region. An observed spectrum from a synchrotron radiation source is thus made up from the assorted spectra emitted by individual electrons and its precise form will depend upon the details of the electrons' energy distribution and the physical structure of the magnetic field. For most astrophysical synchrotron radiation sources though, the intensity (I) decreases as the frequency (ν)

increases in a way that is proportional to a power of the frequency, i.e. $I \propto \nu^{\alpha}$, where α is called the spectral index of the source. The spectral index for emitting regions that are transparent to the radiation usually has a value between −0.2 and −2, giving a reduction in intensity ranging from 40% to 99% for a factor of 10 increase in frequency When the emitting region is opaque, then the spectral index has a value of 2.5, i.e. the intensity increases by a factor of 300 for a 10-fold increase in frequency (Fig. 3.2). Many synchrotron sources emit radiation only at long wavelengths, but in some circumstances the electrons can be sufficiently energetic that x-rays and even gamma rays are produced. Electron energies exceeding 10^{12} eV (Box 1.1) are needed for the production of x-rays, however cosmic ray particles can have energies up to a billion times larger than this, so that the possible existence of sufficiently energetic electrons is quite realistic.

The radiation from thermal and free–free sources is unpolarized when it is produced while synchrotron radiation by contrast is usually strongly linearly polarized (Box 3.2). This difference arises because the magnetic field concerned with synchrotron radiation ensures that the many electrons involved emit their individual photons in alignment with each other. Since free–free emission occurs as a result of individual encounters between electrons and ions, the photons come out with random orientations and there is no net polarization. The polarization difference between synchrotron and other sources is so marked that the observation of strongly polarized radiation is taken to be a certain indicator of synchrotron radiation emission.

Inverse Compton Scattering

This is not a true source of radiation, but a process whereby already existing photons are boosted to higher energies (i.e. to shorter wavelengths and higher frequencies). Compton scattering occurs when x-rays or gamma rays irradiate free electrons in a plasma. Collisions between high-energy photons and low-energy electrons lead to the transfer of energy from the former to the latter. Inverse Compton scattering is, as might be expected, just the reverse of this; during collisions relativistic electrons transfer some of their energy to lower energy photons. The spectrum of the resulting radiation depends upon that of the original radiation source and the energy distribution of the electrons, but is usually similar to that of synchrotron radiation sources (Fig. 3.2) while being unpolarized. Inverse Compton scattering is thought to underlie the production of some of the highest energy x-rays and gamma rays emitted by AGNs.

To return to our attempt to say what is a non-starburst active galaxy, our definition now becomes

"a compact, central, AGN, emitting significant amounts of non-thermal radiation which is contained within a relatively normal host galaxy".

In practice the emissions from the AGN may be so much larger than those from the host galaxy that the latter is swamped and cannot be seen. However as telescopes and observing techniques improve, more and more AGNs, which initially appeared to be isolated, are found to lie inside host galaxies and it seems likely, though not yet proven, that all AGNs are located within more normal galaxies.

In addition to this basic property of AGNs they often, but not always, have some of the following characteristics:

- A very high bolometric[12] luminosity, a luminosity of 10^{37} W (an absolute magnitude of –21 or about two-thirds of the luminosity of the Milky Way Galaxy; Table A5.1 in Appendix 5 provides a handy refer-

[12]The bolometric luminosity is the luminosity given by the total energy emitted over the entire spectrum from the longest radio waves to the shortest gamma rays.

ence for the interconversion of the various luminosity measures used for AGNs), is sometimes taken as the lower limit. While many AGNs are hundreds or thousands of times brighter than this, there is increasing evidence that AGN-type activity can occur at low levels and that there is no sharp distinction between normal galaxies and AGNs. AGNs with a luminosity less than 5×10^{37} W (an absolute magnitude fainter than −23) are sometimes classed as Low-Luminosity AGNs (LLAGN).

- A much higher luminosity than a normal galaxy within one or more of the x-ray, ultraviolet, infrared or radio spectral regions.
- Strong emission lines in the optical spectrum.
- Rapid variability in the luminosity of the AGN.
- Evidence for the ejection of material at high velocities in the form of two opposed narrow jets. The sizes of the jets may range from a few light minutes to millions of light years (10^{-5} pc to 100 kpc). Sometimes only one of the jets may be detectable.
- Pairs of regions or lobes up to 100,000 ly (30 kpc) across that are located up to 1 Mly (0.3 Mpc) away from the AGN and from which strong radio emissions are received. These radio lobes may align with the smaller scale jets if these are to be found.

There is rarely any doubt over the AGN nature of the more extreme examples of these objects and since for several decades these were the main AGNs being studied, it is only recently that the need for a precise definition of an AGN has arisen. The criteria listed above will generally suffice for initial classification, thereafter the more detailed properties of individual types of AGN, discussed in the next section, will come into play. However there remain ambiguities in, for example, deciding if a particular object is a weak Seyfert galaxy or a strong LINER (Low-Ionization Nuclear Emission Region), etc. If though, as discussed above, all galaxies contain super-massive black holes, then the question of whether a particular galaxy is normal or contains an AGN probably becomes less important than might once have seemed possible.

3.2 Types of AGNs

3.2.1 Preview of AGN Models

Once we have established the natures of AGNs and so are able to see what needs explaining, we will look in detail at how the energy in AGNs is generated and how the observed structures within AGNs can originate. The study of AGNs however has developed over many decades and the opening up of new regions of the spectrum from time to time has led to many new and apparently different AGNs being identified. Hence there are now a confusingly large number of seemingly disparate objects all labeled as being AGNs. The achievement of some order within this chaos has come about over the last couple of decades with the development of the unified model for AGNs (Chap. 5). A brief discussion of that model is given here so that the reader can see how the types of AGNs relate to each other and why the order of discussion within the next few sub-sections has been chosen.

The unified model of AGNs starts with a rotating black hole that has a mass of tens, hundreds or thousands of megasuns. This is embedded within a dense disk of material that is falling in towards (being accreted by) the black hole from further out in the galaxy. The extremely high temperatures developed within the accretion disk as gravitational energy is released by the plummeting material may lead to the ejection of matter along the rotation axis of the accretion disk where the surrounding gas is at its thinnest. The ejected material may then be further accelerated to produce relativistic jets. The black hole and accretion disk are surrounded by more clouds of material that may take the form of one or more toruses (i.e. shapes like ring doughnuts) and clusters of smaller gas and dust clouds that hide them from outside observers. All this is then embedded at the heart of a relatively normal galaxy.

The features of the *observed* AGN then depend upon the inclination of the observer's line of sight to the rotation axis. If looking along or near to the rotation axis, then the approaching jet will dominate the emission. If observing more or less perpendicularly to the rotation axis, then most of the AGN will be hidden from direct view, and its energy re-processed by absorption and re-emission within the outer parts of the AGN and the galaxy. The type of AGN that is seen thus depends upon the angle from which we observe it. In addition some AGNs (about 10%) emit strongly at radio wavelengths and are called radio-loud (RLAGN), the rest have radio emissions similar to normal galaxies and are called radio-quiet (RQAGN). Whether or not an AGN is radio-loud or not seems to depend upon the electrons in the jets reaching relativistic speeds. We may thus clarify the AGN confusion somewhat by sorting them into whether they are viewed along or perpendicularly to the accretion disk axis and whether they are radio-loud or quiet (Table 3.1).

3.2.2 Seyfert Galaxies and Related Objects

With hindsight we now know that the earliest identification of AGN-type activity pre-dates by nearly two decades the knowledge that galaxies were independent and gigantic star systems situated far outside the Milky Way Galaxy. Edward Fath of the Lick observatory noted "planetary nebula-type" emission lines in the spectrum of the spiral nebula M 77 (NGC 1068—Fig. 3.4) as early as 1908. While Hubble, as we have seen in Sect. 1.1, did not prove the Andromeda nebula (M 31) to be a star system lying outside the Galaxy until 1924. Similarly the jet in the giant elliptical galaxy M 87 (NGC 4486, Virgo A, 3C274; see Fig. 3.5) – now classified as an FR 1 radio galaxy – was first picked up by Curtis as early as 1918. However the real start of the study of AGNs as an identifiable group of objects dates back to 1943 and to Carl Seyfert (Fig. 3.6)

By the early 1940s, 12 galaxies had been found that, like M 77, had very bright and angularly small nuclei together with spectra containing strong emission lines. At that time Seyfert was employed at the Mount Wilson observatory and decided to study the spectra of six of these in some detail. He found that the emission lines were unusually wide – if their widths were due to Doppler broadening (Box 1.2) then

Table 3.1 Sorting out AGNs.[13]

	Radio-quiet	Radio-loud
Viewed perpendicularly, or nearly so, to the accretion disk's rotation axis	Seyfert galaxy type 2 (Sect. 3.2.2. Also Markarian galaxy, N type galaxy) LINER galaxy (Sect. 3.2.3. Low-Ionization Nuclear Emission Region) NLXG (Sect. 3.2.2. Narrow-Line X-ray Galaxy) QSO type 2? (Sect. 3.2.4. Also Quasi-Stellar Object type 2)	Double-lobed radio galaxy (Sect. 3.2.6. Also Fanaroff–Riley class I (FR I), Fanaroff–Riley class 2 (FR 2)) NLRG (Sect. 3.2.2. Narrow-Line Radio Galaxy) Quasar type 2? (Sect. 3.2.4. Also hidden or buried quasar, x-ray galaxy, some ULIRGs?)
Looking along or nearly along the accretion disk's rotation axis	Seyfert galaxy type I (Sect. 3.2.2. Also Markarian galaxy, N type galaxy) QSO (Sect. 3.2.4. Also Quasi-Stellar Object type I, BALQSO (Broad Absorption-Line Quasi-Stellar Object), Radio-Quiet Quasar, RQQ)	Blazar (Sect. 3.2.5. BL Lac object, OVV quasar (Optically Violently Variable quasar) and High-Polarization Quasar (HPQ)) BLRG (Sect. 3.2.2. Broad Line Radio Galaxy) Core-dominated radio galaxy (Sect. 3.2.6.) Quasar (Sect. 3.2.4. Also Quasar type I, Low-Polarization Quasar (LPQ), Radio-Loud Quasar (RLQ), also see FR 2 objects in Sect. 3.2.6)

velocities of several thousands of kilometers per second were required. The allowed emission lines were produced by hydrogen, while the forbidden lines were due to doubly ionized oxygen and neon and to

[13]Remember, the glossary of galaxies and AGN types (Appendix 2) lists a handy summary of these objects and their properties.

Figure 3.4 M 77 (NGC 1068): AGN emission lines were observed from this galaxy as early as 1908. It is now classed as a Seyfert type-2 galaxy and is the brightest known example of the type. (Image courtesy of the ING archive and Nik Szymanek.)

singly ionized sulphur. Seyfert also noticed that for some galaxies the lines were all of comparable width, while for others the forbidden lines were narrower than the hydrogen lines. The galaxies involved in Seyfert's study were mostly spirals, but one, NGC 1275, clearly was not. Seyfert published his work identifying a new class of unusual objects in 1943. Even before the paper announcing his results appeared in print though, Seyfert had been called away to be employed on war work, so little further was done with his discovery until the 1960s. However this group of galaxies is now called the Seyfert galaxies in recognition of his contributions. From Seyfert's original six galaxies, thousands are have now been found, so that they make up a few per cent of all bright galaxies – perhaps even rising to 10% amongst the

Figure 3.5 An HST image of the jet within M 87. The jet was first noticed in 1918. (Image courtesy of NASA and the Hubble heritage team (STScI/AURA).)

Figure 3.6 Carl Keenan Seyfert (1911–1960) whose work initiated the study of AGNs. (Image courtesy of Vanderbilt University, Dyer Observatory.)

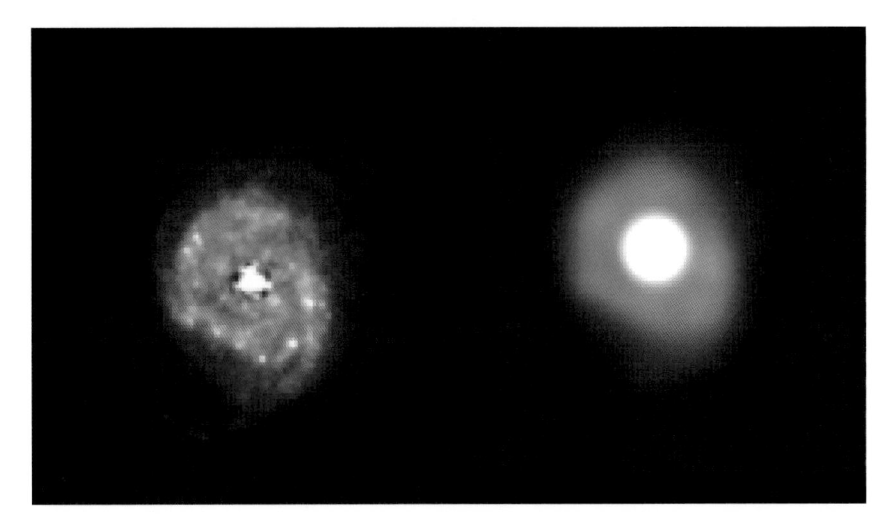

Figure 3.7 A Keck telescope image of NGC 7469, a Seyfert I galaxy. Despite the use of adaptive optics to sharpen the image, the nucleus of the galaxy is unresolved. This galaxy was amongst the first six studied by Seyfert. (Image courtesy of Olivier Lai, David le Mignant and the CFHT AO team.)

weakly active Seyferts. Most Seyferts are found to be spiral galaxies, especially the more tightly wound and the barred types (Sa, SBa, Sb and SBb; see Sect. 1.2.1), but a few reside within irregular galaxies and similar spectra occur from a small number of elliptical galaxies.

Seyfert's original criteria for selecting his galaxies – a small very bright nucleus (Fig. 3.7) and a spectrum containing strong, broad emission lines that came from atoms in a range of ionization stages – still provide the basic definition for the group. However by the 1970s it was becoming clear that the presence of narrower lines in the spectra of some galaxies was of great physical significance and Ed Khachikian and Daniel Weedman sub-divided the group into type-1 and type-2 Seyferts (Fig. 3.8). The hydrogen and other allowed lines in the spectra of Seyfert 1 galaxies are wide, with Doppler broadening velocities of up to 10,000 km/s, while the forbidden lines are much narrower, corresponding to velocities of 1,000 km/s or lower. All the emission lines in Seyfert 2s' spectra by contrast have similar widths that are equivalent to Doppler broadening velocities of 1,000 km/s at most (Fig. 3.8). At the time of writing the numbers of known

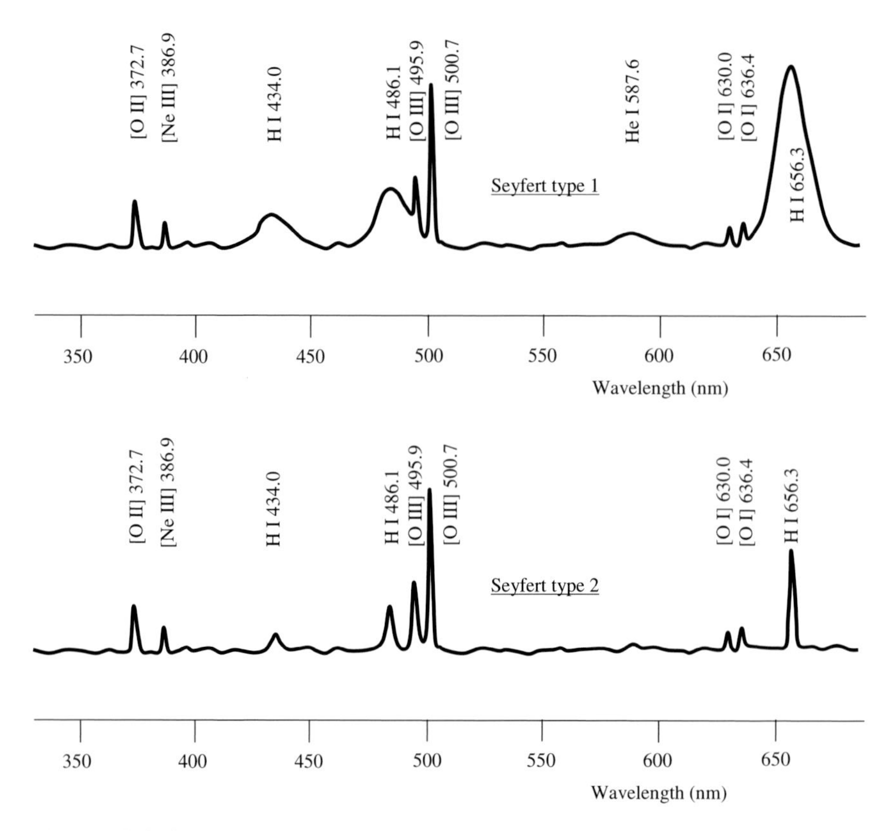

Figure 3.8 Schematic spectra of type-1 and 2 Seyfert galaxies; note the broad hydrogen and helium lines in the Seyfert 1 spectrum (top).

Seyfert 1s and 2s are 9,589[14] and 4,460 (Fig 3.15, below), respectively. However there are another 6,895 objects that are fainter than an absolute magnitude of -23^m that are currently classified as QSOs and quasars, most of which are probably better regarded as Seyfert 1 galaxies. The apparent preponderance of Seyfert 1s over Seyfert 2s may not be quite as large as these figures suggest though, since there is likely to be an observational bias arising from the easier detection of Seyfert 1s. Indeed, some studies

[14]The author is grateful to Dr Mira Véron-Cetty of the Observatoire de Haute Provence for supplying this information in advance of the publication of the 12th edition of the *Catalogue of Quasars and Active Galactic Nuclei* (see Appendix 1).

suggest that the ratio is actually the opposite way round, i.e. Seyfert 2s are two to five times more common than Seyfert 1s. There has been an enormous increase in the number of recognized AGNs very recently (for example just 2,765 Seyfert 1s were known in 2001) as a result of the Anglo-Australian Telescope's (AAT) 2dF galaxy redshift survey and the Sloan Digital Sky Survey (SDSS, see Box 3.2).

Box 3.2 Galaxy Searches and Surveys

The more-or-less chance discovery of AGNs that was the norm during the early stages of their study has been replaced by systematic searches. As mentioned in the main text, Benjamin Markarian using the 1.3-m Schmidt camera of the Byurakan observatory in Armenia together with an objective prism had, by 1981, found some 1,350 starburst and 150 Seyfert galaxies.

During the late 1990s, the 3.9-m AAT was equipped with corrector lenses that gave sharp images over a field of view two degrees across (2dF). Using a spectroscope fed by fiber optic cables, spectra of up to 400 galaxies could then be obtained from within this field with just a single exposure. A survey of the redshifts of some 220,000 galaxies (Galaxy Redshift Survey, GRS) was undertaken using this equipment between 1997 and 2001, discovering some 10,000 QSOs and quasars plus many other types of AGNs in the process.

The SDSS aims to survey a quarter of the whole sky using a purpose-built 2.5-m telescope at Apache Point Observatory in New Mexico. Like the 2dF project, the SDSS uses a multi-optical fiber feed for its spectroscope, enabling 640 spectra to be obtained simultaneously. By the time that the survey is completed (which is expected around the time of writing of this book) it should have measured the positions and brightnesses of 100 million objects and obtained the spectra of a million galaxies and 100,000 QSOs and quasars. With the fourth data release in 2005, 76,483 QSOs and quasars had been identified.

AGNs of many types are also to be found amongst the results of other surveys such as the IRAS catalogue of over a quarter of a million galaxies observed in the infrared, 2 MASS (Two Micron All Sky Survey: data on one million galaxies based upon observations of two 1.3-m telescopes in Arizona and Chile) and 6dF (Six Degree Field: a spectroscopic survey using the 1.2-m UK Schmidt camera at the Anglo-Australian Observatory (AAO) and covering 150,000 galaxies), etc. X-ray surveys also tend to pick up AGNs since they often emit strongly at short wavelengths, but further details will be left for the interested reader to research for him or herself.

A regularly up-dated list of the known Quasars and AGNs is maintained at the astronomical data center in Strasbourg (Centre de Données Stellaires, CDS). The twelfth edition of the *Catalogue of Quasars and Active Galactic Nuclei* by Drs M.-P. Véron-Cetty and P. Veron is currently being compiled and will contain over 100,000 entries – nearly 85,000 just for QSOs and quasars alone (Fig. 3.15, below). It, when it is ready, and the 11th and earlier editions are only available electronically and are open to any enquirer at http://cdsweb.u-strasbg.fr. Further access details are given in Appendix 1. A useful but incomplete list of catalogues and surveys that include AGNs may be found in Allen's *Astrophysical Quantities* (Appendix 1).

In 1981 Donald Osterbrock refined the classification further by introducing the sub-types 1.2, 1.5, 1.8 and 1.9 on the basis of the hydrogen line shapes. In these transitional types, the hydrogen emission lines have a narrow component superimposed upon a much wider component. The relative strength of the narrow component in comparison with the wide component increases along the sequence from 1.2 to 1.9, with Seyfert type 1.5s being intermediate between types 1 and 2. Minor refinements have been added to the classification since then, but Osterbrock's system is currently still in widespread use. In a few cases the Seyfert type of an individual galaxy has been known to change over a period of time, sometimes by as much as between types 1 and 1.9.

Narrow-Line Seyfert 1 galaxies (NLS1s) were added as a sub-class of the Seyfert 1s by Osterbrock in 1985. Their broad emission lines are less than 2,000 km/s in width but the ratio of the strengths of the [O III] 500.7 and H I 486.1 spectrum lines is lower than in Seyfert 2 galaxies and there are other differences such as strong emission lines from ionized iron that distinguish them from Seyfert 2s. It is possible that the differences from normal Seyfert 1s arise from these galaxies hosting central black holes with lower than usual masses, but this is not certain.

Although they are wider than the lines in the spectra of classical galaxies, in the context of AGNs, the forbidden lines in Seyfert 1s and all the lines in Seyfert 2s are classed as "narrow" lines, whilst the wider allowed lines in Seyfert 1s are called "broad" lines. The speeds of the material emitting Seyfert 1s' broad lines are clearly much greater than those of the material producing the narrow lines, so that there must be two distinct and separate emission-line regions within these galaxies. The similar widths of the lines in Seyfert 2 galaxies suggest that just a single emitting region is involved. The two emitting regions are often called the Broad Line Region (BLR) and the Narrow-Line Region (NLR) respectively. The separation of Seyfert galaxies into types 1 and 2 is sometimes extended to other types of AGNs, with the spectra of type-1 objects containing BLR emissions while type-2 objects contain only NLR emissions. Thus QSOs and quasars are type-1 objects, but it is expected that type-2 versions of these exist and that they may take the form of some ULIRGs, radio galaxies and x-ray galaxies (see below and Sects. 2.2 and 3.2.4).

The apparently clear criterion that Seyfert 2 galaxies do not contain any BLR emissions has become muddied in the last decade. When they are observed in polarized light (Box 3.3), the spectra of some Seyfert 2 galaxies turn out to be similar to those of Seyfert 1 galaxies. Since the intensity of the polarized component is only a small fraction of the total, the BLR emissions are swamped when the full spectrum is observed. The polarized radiation results from light that had not originally been directed towards the observer but which has been directed into the line of sight by scattering. We have already seen in the preview of the unified AGN model (Sect. 3.2.1) that Seyfert 1 galaxies are thought to have the directions of their accretion disks' rotation axes pointing close to the observer, while Seyfert 2 galaxies are probably seen from adjacent to the "equators". Thus it seems likely that these Seyfert 2 galaxies are emitting Seyfert 1-type spectra in directions near to their rotation axes and that some of that light is then being scattered towards the observer by material far from the centers of the galaxies. To complete the picture we may hence infer that both the BLR and NLR emissions may be seen from positions near to the rotation axes, but only the NLR emissions from positions at right angles to this. The NLR thus must be large (perhaps 30 ly to 3,000 ly across or more – 10 pc to 1 kpc) and be located outside the BLR, which may be a small as 3 ly (1 pc) in size and perhaps much smaller than that (see below). Something then lies between the two regions and absorbs the BLR emissions in the equatorial plane, but not in the directions towards the poles – the "something" could perhaps be a thick dust torus (Chap. 5). Whether we can go on from this to conclude that all Seyfert galaxies are actually type 1s, and only the viewing angle causes some to appear as type 2s, is still an unresolved question and there remains the possibility that some, perhaps many, Seyfert 2 galaxies genuinely do not host BLRs. Some evidence to suggest that Seyfert 2s do differ from Seyfert 1s by more than just the angle at which they are seen comes from recent HST observations indicating that their host galaxies contain more dust and are of later Hubble types (Sect. 1.2.1) than those of Seyfert 1s. But it may be that such host galaxies are just more likely to hide their BLRs and so appear as Seyfert 2s.

Box 3.3 Polarization

The polarization of light will be familiar to many people from using their Polaroid™ sunglasses. However e-m radiation of any wavelength may be polarized and the polarization comes in two varieties.

In linear polarization, the waves all have their vibration directions aligned with each other (in unpolarized light the directions of the vibrations are distributed at random). In elliptical and circular polarization, the vibration directions are again aligned, but the orientation of that alignment rotates at the frequency of the radiation. Naturally produced polarized radiation is a mixture of unpolarized, linearly and elliptically polarized radiation. The degree of polarization is the ratio of the intensity of the polarized component(s) to the total intensity. The degree of linear polarization can reach 60% or more from some synchrotron sources (Box 3.1) and ranges from that down to the limits of measurement. Astrophysical sources are rarely elliptically polarized and for the few exceptions to this, the degree of elliptical polarization is small.

In astrophysics, polarized radiation arises in three main ways – either through the reflection or scattering of existing unpolarized radiation or because synchrotron emission is the source of the radiation. Synchrotron radiation is discussed in Box 3.1. Polarization by reflection can easily be demonstrated by looking at the reflections off (say) a car windscreen through Polaroid sunglasses. If you rotate the sunglasses (take them off your head to do this!), then when the planes of polarization of the sunglasses and the reflected light are mutually perpendicular, you will find that the reflections almost disappear. Similarly the blue light from a clear sky is scattered sunlight and is slightly polarized. The maximum degree of polarization occurs at 90° to the position of the Sun in the sky and observing through a rotating pair of Polaroid sunglasses will show that area of the sky lighten and darken as the polarized component is either allowed through or rejected by the Polaroid (take care not to look anywhere close to the Sun if you try this out).

The emitted radiation from AGNs is often polarized and both scattering and synchrotron radiation contribute to the effect. The existence and nature of the polarized emission often reveals far more about the nature of AGNs than comes from studying the unpolarized radiation.

The optical luminosity of the brighter Seyfert galaxies is dominated by the emission from the nucleus. On short exposures, only the star-like core of the nucleus will be visible. Very distant Seyfert galaxies likewise may only show the AGN and then they may come to resemble low-luminosity QSOs and quasars (Sect. 3.2.4). A fairly arbitrary borderline is defined at an absolute bolometric magnitude of -23^{m}; objects brighter than this are considered to be QSOs, quasars, blazars, etc. (see subsequent sections), objects fainter are taken to be Seyferts, LINERS (Sect. 3.2.3) and the like. Hence the brightest Seyferts are a few times more luminous than the Milky Way Galaxy.

The faintest Seyferts have only a fraction of one per cent of the luminosity of the Milky Way Galaxy. Thus the dwarf Seyfert 1 galaxy, NGC 4395, which is the faintest known Seyfert (and also the nearest Seyfert to us at a distance of 8 Mly (2.5 Mpc)) has an absolute bolometric magnitude for its nucleus of just -10^{m}. The fainter Seyferts overlap with the LINERS in many of their properties and the two groups may be difficult to distinguish observationally. One useful criterion in such cases is based upon the ratios of the intensities of some of the forbidden emission lines to those of the allowed lines and was developed in the 1980s by Jack Baldwin, Mark Phillips and Roberto Terlevich amongst others. Their method serves as a means of separating Seyferts and LINERS and also helps to distinguish both from H II regions. The lines of doubly ionized oxygen and of singly ionized nitrogen compared with the hydrogen line intensities are one widely used example of this method (Fig. 3.9). However it is still unclear whether LINERs are different in kind from Seyferts or whether there is a smooth continuum from classical galaxies through LINERs to the Seyferts (and as we shall see, perhaps then continuing on to the QSOs and quasars).

Seyferts are thus intermediate between LINERS (Sect. 3.2.3) and QSOs in their levels of activity. The optical luminosities of Seyferts are similar to those of normal spiral galaxies, but they also emit additional energy in the x-ray, ultraviolet and especially the infrared regions. In the latter case the radiation is probably thermal emission from hot dust particles. Many Seyferts have an emission line due to iron in the x-ray

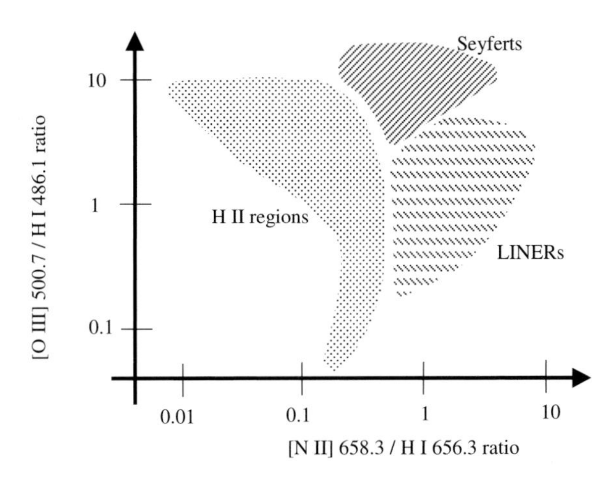

Figure 3.9 A schematic Baldwin–Phillips–Terlevich plot using the line ratios [O III] 500.7/H I 486.1 and [N II] 658.3/H I 653.6 showing the separate areas occupied by Seyferts, LINERS and H II regions.

region at an energy of 6.4 keV (a wavelength of 0.2 nm). Known as the Kα line, it is sometimes very broad and double peaked and it has been suggested that it originates directly from the accretion disk around the central black hole (Sect. 3.2.1). Almost all Seyferts are radio-quiet, the few exceptions usually deviating from the standard pattern in other ways as well. There is evidence that there is some correlation between the luminosities of the AGN and of the host galaxy – the brightest AGNs tend to reside within the brightest galaxies – although recent observations made using the Gemini telescope have not upheld this relationship. Also there is a link with the emission linewidths, these being likely to be greatest amongst the most luminous galaxies.

Maser emission (Box 3.1) has been detected coming from a number of galaxies in recent years, including AGNs. These masers can be over a million times more luminous, up to 10^{30} W, than masers within our the Milky Way Galaxy and so have become named "megamasers". The masers are mostly due to hydroxyl (OH) and water (H_2O) molecules, with the OH masers predominantly occurring within ULIRGS (Sect. 2.2) and the H_2O masers within Seyfert 2 and LINER galaxies (Sect. 3.2.3). The H_2O masers are found clustered very close to the centers of the AGNs and their alignments and velocities suggest that they occur within thin molecular disks around the central black holes. The almost complete absence of observed H_2O maser emission from Seyfert 1 galaxies suggests that the emission is preferentially beamed perpendicularly to the central accretion disks' rotation axes and this is discussed further in Chap. 5.

The spectra of normal galaxies result from the combined spectra of their component stars, so they have continuous spectra with superimposed absorption lines. In Seyfert galaxies, similar underlying spectra may be detectable, but there is also continuous emission that lacks any absorption features extending from the blue part of the spectrum through, sometimes, to the x-ray and gamma-ray regions as far as 100 keV (0.01 nm). Often there is a peak in the continuum energy emission in the ultraviolet that in many cases extends into the blue part of the visible spectrum. Known as the Big Blue Bump (BBB) it may be thermal emission from the accretion disk. The blue continuum is strongest for the Seyfert 1s and weaker, sometimes absent, in the Seyfert 2s and was the identifying feature sought by Markarian during his surveys (see below) that enabled him to pick out Seyfert galaxies from other types.

The AGNs in some Seyferts vary in brightness with time, typically taking a few months for significant changes to occur, though their x-ray luminosities can alter in matter of a few tens of minutes. This limits the sizes of the emitting regions (Box 3.4) to being less than a few light-months ($<$0.1 pc) and to being something comparable with the size of the inner solar system for the x-ray producing regions.

The variability of AGNs can be further exploited to help understand their structures through reverberation mapping (Box 3.5). While there are many uncertainties in this process, time delays of up to 10 days are found for the broad allowed lines from highly ionized atoms and about twice that for the lines from less highly ionized and neutral atoms. The sizes of BLRs are thus probably in the region of a few light days. NLRs are much larger than BLRs and can sometimes be resolved directly by ground-based telescopes using active atmospheric compensation or by the Hubble space telescope. They are found to range from a few tens to a few thousands of light years across (10 pc to 1 kpc). For some galaxies the NLR can stretch out even further out from their centers – to 50,000 ly (20 kpc) or more – they then become called Extended Narrow-Line Regions (ENLRs).

Box 3.4 Light Size

The speed with which an object emitting radiation brightens or fades by a significant amount sets a limit on that object's physical size. The limit, known as the Light Size, is the distance traveled by electromagnetic radiation in the time taken for the object's brightness to vary. It is about the maximum possible physical size for the emitting region. Although some ingenious morphologies and behavior patterns can be devised for the emitting region to allow it to exceed the light size by small amounts, these are not very realistic. Super-luminal motions (Chap. 4) however can compress the times taken for brightness changes to occur by a similar factor to the excess of the observed speed over the speed of light. Some care thus needs to be taken if using light size when studying Blazars (Sect. 3.2.5), etc.

More importantly, rapid brightness changes can occur that seem to imply that the object is much smaller than its real size. Thus a supernova occurring within a distant, unresolved, galaxy, could suggest a light size of just a few light days (around 100,000 million kilometers), whereas the galaxy may actually be 100,000 ly (30 kpc) across. Provided that the physical situation is reasonably well understood, however, such ambiguities should be rare. In the just quoted example, a spectrum, or even the shape of the light curve, is likely to reveal the emitting region to be a supernova, not the whole galaxy. The light size theorem is then again valid, since the extent of a supernova is a few tens of millions of kilometers – far smaller than the light size suggested by its brightness changing over a few days.

With active galaxies, the light size theorem is used extensively to put upper limits on the sizes of their emitting regions. 3C48, the first quasar to be found, was found to be variable on a timescale of a year soon after its discovery. Its emitting region must thus be smaller than about one ly (0.3 pc). It also varied daily, but only by 1 or 2%. This is too small a variation to constrain the size of the main emitting region, but it does show that there must be much smaller emitting regions within the main one. The optical emissions from other quasars, QSOs and Seyferts, particularly Seyfert 1s, vary over periods as short as a few days. At radio wavelengths some AGNs are constant, at least on timescales of a few tens of years, even though they may change their brightnesses at shorter wavelengths. Others may vary rapidly – the quasar PKS 0405–385 has undergone significant radio luminosity changes in less than an hour. The speed records for variability however occur at x-ray wavelengths, there, most AGNs change luminosity over a few hours, a few days or a few weeks. One, quasar PKS 0558–504, has even brightened by a factor of 67% in just three minutes. The size of its x-ray emitting region must thus be less than the size of Mercury's orbit.

The derivation of the light size theorem is illustrated in Fig. 3.10. An observer monitors the brightness of an object one Mly (300 kpc) away. Supposing the object to be opaque (if it is translucent, then its size is even more constrained) and about two ly (0.6 pc) across, then the observer will receive radiation from a depth of about one ly (0.3 pc). Suppose now that the object instantaneously doubles in brightness all over its visible surface. Normally, of course, any change in brightness will take longer and occur at different times and possibly by differing amounts for different parts of the surface. However something close to such an instant change is possible, for example if a supernova were to explode at the center of a spherical gaseous nebula, then the increase in its emitted energy would reach the various parts of the visible surface of the nebula simultaneously. The increased emission from the nearest portion of the object will reach the observer one million years later (this is a rather long-lived

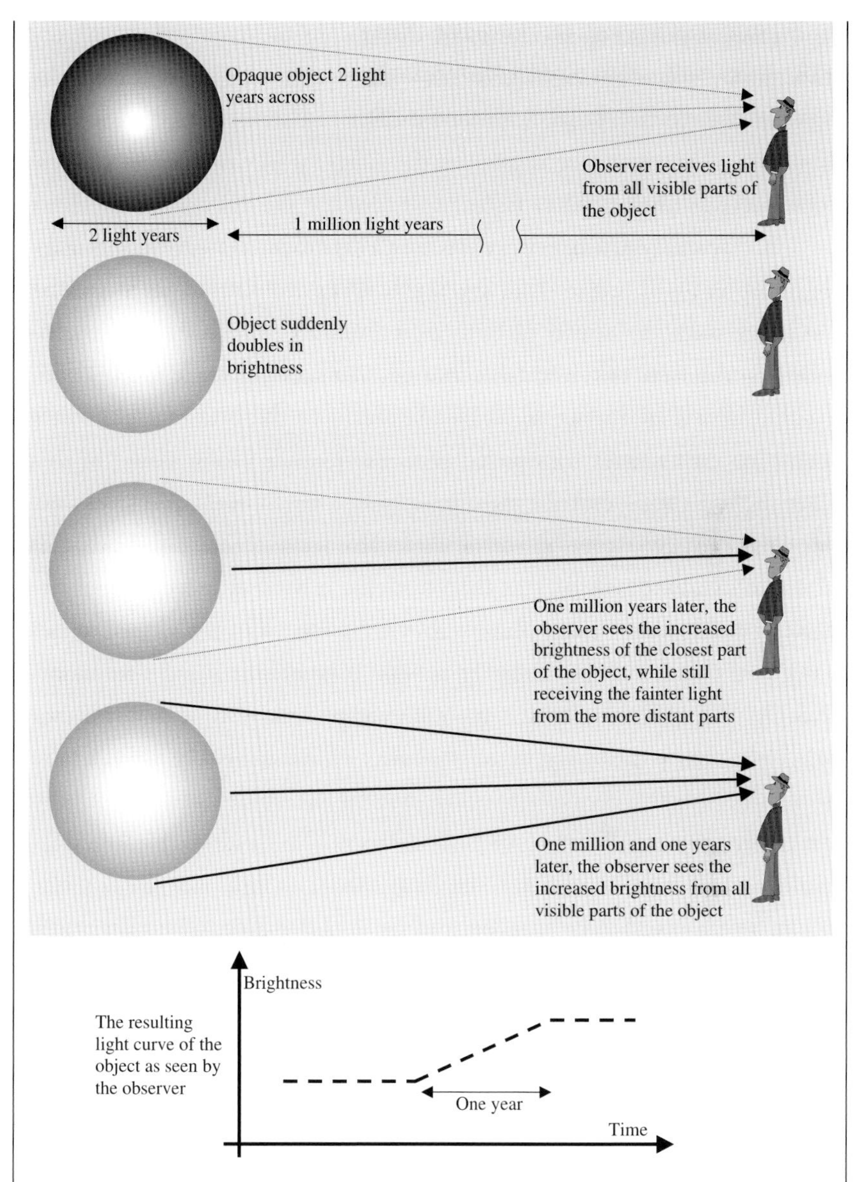

Figure 3.10 The light size theorem: A significant variation in brightness over one year constrains the maximum size of the emitting region to be about one ly.

and patient observer!). The most distant visible parts of the object though are a further light year (0.3 pc) away and so it will take the increased radiation from there one million and one years to reach the observer. The observer thus sees the original instantaneous change smeared out until it takes a year to occur.

Conversely an observer detecting a significant brightness change in an object will realize that its depth along the line of sight can be no more than the speed of light in a vacuum (300,000 km/s) multiplied by the time taken for the change. Objects can be long and thin, so that if oriented near to perpendicularly to the line of sight, their depth could be small compared with their other dimensions. In this case the light size could be significantly smaller than the physical size. However given a random orientation for such objects, the error in the size estimate will be less than a factor of two at least 67% of the time. Furthermore, most objects that we observe in the skies are not long and thin but have roughly similar dimensions in all directions. Finally, the uniform, instantaneous change assumed in determining the light size is unlikely to be realistic. More probably it will take some time for the changes to occur within the object, and when they do occur they are likely to be non-uniform. Thus, although it is possible for an emission region to be larger than its light size, in practice it can be expected to be smaller, and most of the time to be very much smaller than that limit.

Box 3.5 Reverberation Mapping

The concept of reverberation mapping is straightforward and closely related to the idea of light size. In practice, though, using reverberation mapping is fraught with difficulties. Figure 3.11 illustrates the principle of reverberation mapping. There is a distant object that has a central energy source hidden from the observer. There are several emission regions powered by the central energy source that are at different distances from it. The object is angularly unresolved so that the emission regions cannot be seen separately. The emission regions are all at the same radial distance from the observer and can be seen directly. If the central energy source now increases its brightness then, since it is hidden from the observer, no immediate change will be seen. However if, say, there is an emission region one light day away from the energy source then that region will receive an increased input of energy and so brighten itself one day after the change has occurred in the central energy source. The observer will then see the object increase in brightness. There will be another observed increase in brightness a day later when a second emission region, two light days out from the energy source, receives its increased energy supply. Finally in the situation envisaged in Fig. 3.11 a third increase will follow after a further two days as an emission region four light days out brightens.

Reverberation mapping attempts to use the time delays between successive changes in the emission from an object, which all result from a single change in the intensity of the energy source, to infer the physical structure of the object. For the idealized situation pictured in Fig. 3.11, this is clearly a simple task. In reality the emission regions are likely to be at different distances from the observer so that the direct relationship between the time delay and the distance of the emission region from the energy source is lost. For example in Fig. 3.11, if the emission region that is one light day out from the energy source were to be (almost) behind the energy source, then its changed brightness would be seen on day 3, not on day 1. Secondly the emission regions may themselves be large enough to require times comparable with the delays between emission regions for their brightness changes to occur. In particular, in AGNs spherical or toroidal regions surrounding the central energy source are likely to be found. The observed brightness changes are thus likely to be smeared-out and the change due to one emission region may run into the change from another.

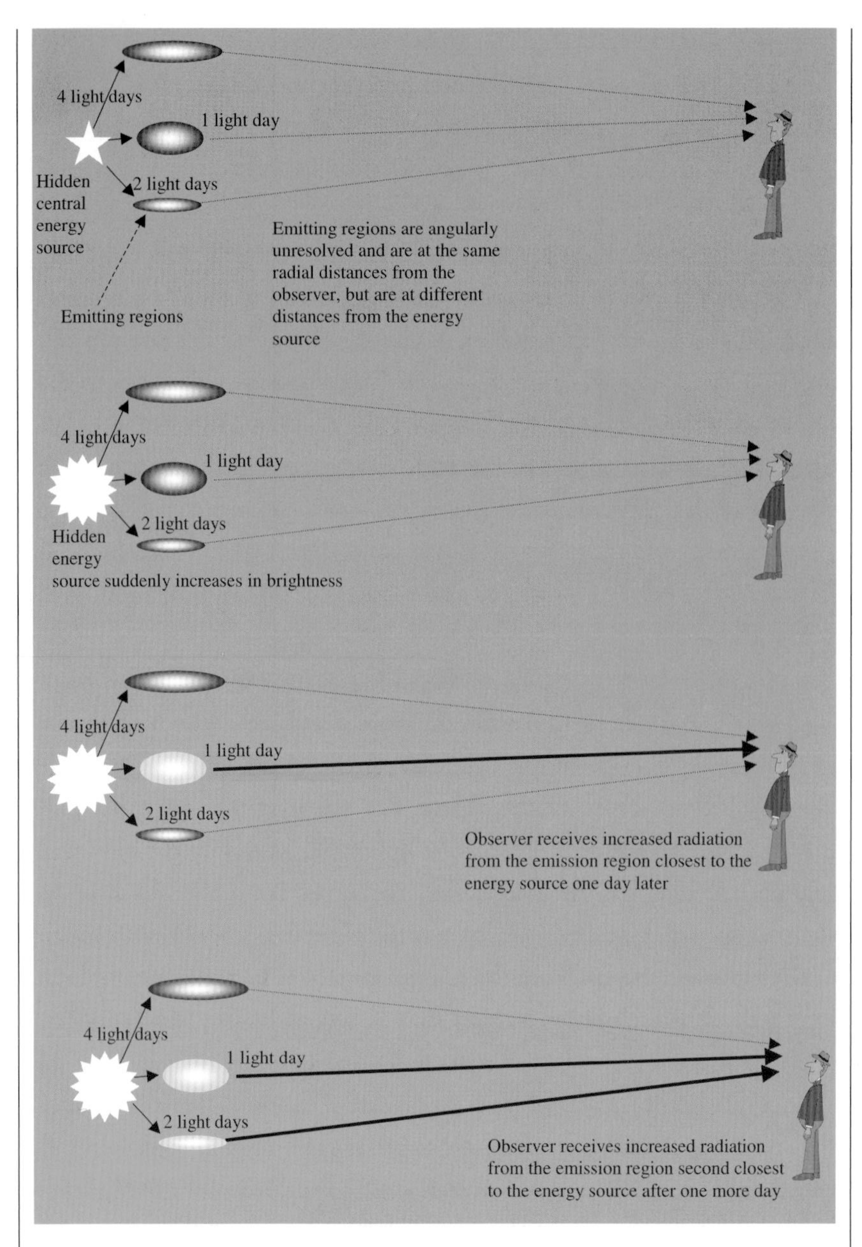

Figure 3.11 Time delays between emission region changes.

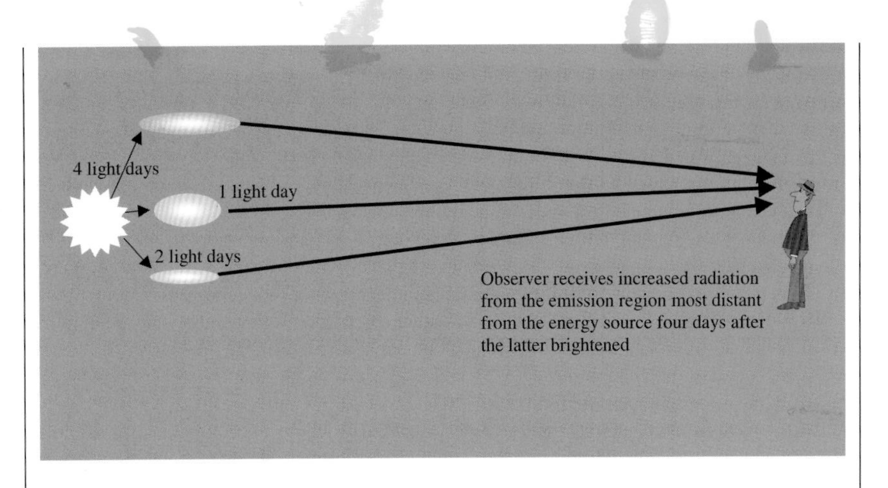

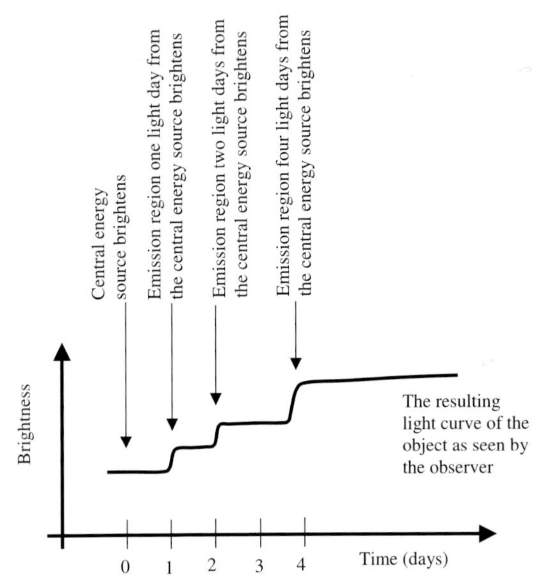

Figure 3.11 *Continued*

One bonus for the reverberation mapping of AGNs however is that time delays between the changes in individual emission lines or groups of emission lines may be monitored, rather than changes in the total luminosity of the source. Overlapping changes can thus be identified and the method used, for example, to study the structures within BLRs separately from those of other parts of the AGN. A crosscheck on the reality of structures obtained via reverberation mapping is that, if the unified model of AGNs (Sect. 3.2.1) is correct, temperatures should be highest in the regions closest to the central black hole. The lines deriving from highly ionized atoms should thus suggest smaller sized regions than those coming from lower ions or neutral atoms.

Finally we may mention some further types of AGNs that seem to be related to the Seyferts or at least have some features in common with them. N-type galaxies were identified early on in the development of research into AGNs. They are now recognized to be distant Seyfert galaxies, and the class is no longer much used. Narrow-line radio galaxies (NLRGs) and narrow-line x-ray galaxies (NLXGs) are galaxies whose spectra are generally very similar to those of Seyfert 2s but which emit strongly at radio and x-ray wavelengths respectively. The brighter NLRGs may be hidden (type-2) quasars (Sect. 3.2.4). Broad-line radio galaxies (BLRGs) may similarly be the radio-loud equivalent of Seyfert 1 galaxies. NLRGs and BLRGs are also sometimes labeled as types of radio galaxies (Sect. 3.2.6). They tend to be associated with elliptical host galaxies rather than the spirals of most Seyferts, but the reason for this is not known.

Markarian galaxies (abbreviated Mrk or MKN) are not a different type of AGN. They are galaxies discovered during a survey conducted by Benjamin Markarian between 1962 and 1981. The survey used the 1.3-m Schmidt camera of the Byurakan observatory in Armenia together with an objective prism to find galaxies with strong emission in the blue region of the spectrum. Of the 1,500 galaxies so discovered, about 10% were Seyfert galaxies and the remainder mostly starburst galaxies. The search however had an observational bias, since by looking for a strong blue continuum, Markarian preferentially selected Seyfert 1 galaxies over Seyfert 2s.

3.2.3 Low-Ionization Nuclear Emission-Line Region Galaxies (LINERs)

Surveying bright galaxies in 1980, Tim Heckman found that a large proportion (we now estimate 30–40%) had weak emission lines in the spectra of their nuclei. Since the atoms emitting the lines were neutral or in low-ionization states the galaxies were rapidly named LINERs (see the section title for the full version). Like Seyfert galaxies, LINER spectra contained a strong emission line due to neutral oxygen at 630 nm, but lack, or only have present weakly, the strong lines due to doubly ionized oxygen and quadruply ionized neon that occur in Seyfert spectra. LINER spectra also have emission lines due to hydrogen. The lines are narrow in the AGN sense, but sometimes weak broad components can be found. By analogy with the Seyferts, such galaxies are then classed as LINER 1s, while those with no sign of broad lines are LINER 2s. LINERs are radio-quiet with about the same radio luminosity as normal galaxies. Their radio spectra however are frequently flat, enabling them to be distinguished from normal galaxies. They may also emit x-rays more strongly than normal galaxies from the centers of their nuclei.

LINERs are also similar to Seyferts in that they occur predominantly amongst spiral galaxies and especially amongst the early Hubble types (Sa, SBa, Sb and SBb), although a small proportion are to be found in peculiar galaxies. LINERs' level of activity places them between starburst galaxies and Seyferts, but whether they are mini-Seyferts or mega-starbursts is still unclear – and of course it could be that both types of activity can lead to LINERs. Up to a fifth of LINERs have compact central ultraviolet or radio sources that suggest that these galaxies at least plus the LINER 1s are related to Seyfert galaxies. Furthermore, although Seyferts and LINERs occupy different regions on Baldwin–Phillips–Terlevich plots (Fig. 3.9), the two regions run into each other without a clear separation thus hinting that there is indeed a continuity of types.

A number of suggestions to explain the energy source for the emission lines have been made, but again which, if any, is correct remains uncertain. The leading contenders are photo-ionization by ultraviolet radiation from a central, non-stellar source (almost certainly this would involve a massive black hole like "normal" AGNs), the energy from massive star-forming regions (starbursts) or from groups of very hot evolved stars and the heating of the gas by shock waves that perhaps originate from supernova explosions. It seems likely that in some cases more than one energy source may be involved – M 58 (Fig. 3.12), for example, contains both a central LINER and a larger starburst region.

Figure 3.12 M 58 (NGC 4579), in Virgo. A possible LINER galaxy some 60 million light years (20 Mpc) away from us. (Image courtesy of Johan Knapen and Nik Szymanek.)

3.2.4 Quasi-Stellar Objects (QSOs) and Quasars

Twenty years after Seyfert identified the first AGNs another and highly spectacular type was discovered but it would be a long time before they were recognized for what they were. Under the pressure of military needs during the Second World War – especially for radar – the equipment and techniques for receiving and transmitting radio waves had developed rapidly. After the end of the war, a great deal of surplus material was sold off, including large radio dishes. Academics returning to their universities after war service took advantage of this to set up radio telescopes for astronomical purposes, with the universities of Manchester (Jodrell Bank) and Cambridge (Mullard Radio Astronomy Laboratory) leading the way in the UK. Radio astronomy was the first of the "new" astronomies and has subsequently been followed by the opening up of the infrared, ultraviolet, x-ray and gamma-ray regions of the spectrum to astronomical observation. Whenever a new segment of the spectrum becomes available in this way, one of the first tasks for observers is to conduct a search to "see what's there". In the 1950s therefore workers at the Mullard Radio Astronomy Laboratory started conducting a series of sky surveys. To begin with the results were patchy – many sources were seen, but some then turned out to be artefacts arising from defects in the equipment. However by the time of the third survey[15] in 1959, which was conducted at a wavelength of 1.9 meters (159 MHz), these problems had been sorted out and the resulting catalogue of nearly 500 radio sources was reliable.

Many of the 3C (third Cambridge) sources were quickly identified with supernova remnants or, with rather less certainty, as galaxies (radio galaxies; see Sect. 3.2.6). However despite a small interferometer being

[15]Sources in this catalogue are designated as "3C" followed by a numeral in right ascension order, for example; 3C31 and 3C437. Sources added later to the catalogue in subsequent revisions are labeled 3CR and given a decimal number; 3CR175.1 thus has an RA placing it between 3C175 and 3C176.

used for the observations, the resolution of the radio telescope was low. This meant that some of sources were unresolved (i.e. they were indistinguishable from point sources and hence called at the time "radio stars"). The low resolution also meant that in such cases the error boxes (Box 3.6) were several minutes of arc across, so that identifying such a radio "star" with its optical counterpart was difficult and doubtful – when it was not completely impossible.

Box 3.6 Error Boxes and Identifications

The resolution of any telescope operating at any wavelength is limited by the wave nature of e-m radiation (Box 1.1). The restriction on resolution imposed by the nature of radiation is called the diffraction limit (because it is diffraction at the edges of the optical components that constrain the sharpness of the images) or sometimes the Rayleigh criterion after Lord Rayleigh who first defined it. The Rayleigh criterion is an arbitrary measure that in practice usually turns out to be close to the actually achievable resolution. If the telescope has a circular aperture, then in the optical region the Rayleigh criterion gives its resolution as about a tenth of a second of arc divided by the telescope's aperture in meters, i.e. a 0.1-m telescope has a resolution of 1", a 1-m telescope a resolution of 0.1" and a 10-m telescope a resolution of 0.01". The same considerations apply within other parts of the spectrum, but the resolution worsens in direct proportion to the increase in wavelength, or improves if the wavelength decreases. Thus at longer wavelengths for example, a 100-m diameter radio telescope operating at a wavelength of 0.1 m would have a resolution of about four minutes of arc, 2,500 times worse than that of a 1-m optical telescope. This extreme blurring explains why radio astronomers usually combine the outputs from several well-separated individual radio telescopes into interferometers with much improved resolutions. Other effects, such as blurring caused the Earth's atmosphere and less than perfect optical components, are likely to make the resolution poorer than that predicted by the Rayleigh criterion.

This limit on the resolution of even the best and largest telescopes means that there is also a similar limit to the precision with which an object's position in the sky can be measured using those telescopes. It is thus customary to present measured positions, not as precise points, but as having a certain probability of lying somewhere within an area whose shape and size are determined by the telescope's resolution. The potential position area is called the error box of the telescope and while it is often circular or square in shape, this is not always the case. An interferometer, for example, may have a better resolution in one direction than another, making its error box rectangular or elliptical.

When an observer has obtained a position for an object by observations at one wavelength, it is common to want to see how that object appears at another wavelength. Often, though not always, this means finding the optical counterpart of a radio or x-ray source, etc. (i.e. "identifying" it). Given the number of objects now detectable on optical images from large telescopes, the error box from the radio or x-ray observations needs to be no more than a few seconds of arc across if just one optical object is to be pinned down. Larger error boxes than this are likely to include many optical objects so that no certain identification of the radio or x-ray source can be made. There is one exception to this rule – if there is one really unusual source within the error box, such as a supernova remnant or a peculiar galaxy, then that unusual object may be taken as the preliminary identification. However it should always be remembered that such an identification is no more than an educated guess and that the radio or x-ray source may turn out to be one of the other, less apparently exciting objects, within the error box.

One of the brighter radio "stars", 3C48, caused a good deal of interest because interferometer measurements at Jodrell Bank had shown that it was under one second of arc in size. In 1960 its position in Triangulum was pinned down to within about five seconds of arc through observations made using the interferometer at the Owens valley radio observatory in California. This instrument was based upon two 26-m dish aerials that could operate at wavelengths as short as 0.3 meters (1 GHz). The aerials could be moved along both east–west and north–south tracks so that, unlike most other interferometers at that time, it provided high resolution (and so position) along two axes. The accurate position for the radio source quickly enabled Alan Sandage and Jesse Greenstein to identify it with a 16^m blue star surrounded by a faint nebulosity. A spectrum then revealed numerous strong and broad emission lines. Emission-lines in stellar spectra are very rare, but few types of stars – such as the Wolf–Rayets – do have strong emission lines, so 3C48 could still be a star. However the lines in 3C48's spectrum had another peculiarity – they did not seem to be identifiable with any of the known elements. John Bolton, Director of the Owens valley radio observatory, suggested a fit to the known elements was possible if the lines had a redshift (Box 1.2) of 0.37. However at that time the largest known redshifts were much smaller, so the suggestion was not taken any further (though a year later Rudolph Minkowski obtained a record redshift of 0.46 for the radio galaxy 3C295).

For three years 3C48 remained an enigma, though still thought by most astronomers to be some type of star. The clue to its true nature came only when a second similar "star" was found. This was another of the Cambridge radio sources, 3C273 in Virgo. It was also identified with an optical object after an accurate position for the radio source was obtained. This time however the position was obtained quite differently from that of 3C48. In 1962, the Moon was due to occult (pass in front of) 3C273 several times. Cyril Hazard then working with the 64-m Parkes radio telescope in New South Wales, Australia, managed to time two of the occultations, even though it meant removing some of the panels from the telescope in order to observe sufficiently close to the horizon. At the instant that an object disappears (or reappears from) behind the Moon it must be located somewhere along the leading (or trailing) edge of the Moon in the latter's motion across the sky. Now the Moon's position is known with high accuracy and so the observation of a single disappearance or reappearance pins down the object's position to somewhere along the arc given by the Moon's edge at that instant. A second observation will give a second and different arc that will intersect the first at one or two points. The object must then be at the position of that point or at one of the two points if two intersections occur. A third occultation will then sort out which of the two points is the right one – or it may be that the radio position will be sufficient for this purpose.

Hazard published a position for 3C273 in 1963 that had an uncertainty of just one second of arc, and also revealed the radio source to be double. A 13^m star coincided with one of the sources and a much fainter jet with the other (Fig. 3.14). In late 1962 Maarten Schmidt (Fig 3.13) used the 5-m Hale telescope on Mount Palomar to obtain a

Figure 3.13 Maarten Schmidt who, along with Cyril Hazard, solved the mystery of the quasars. (Image courtesy of California Institute of Technology.)

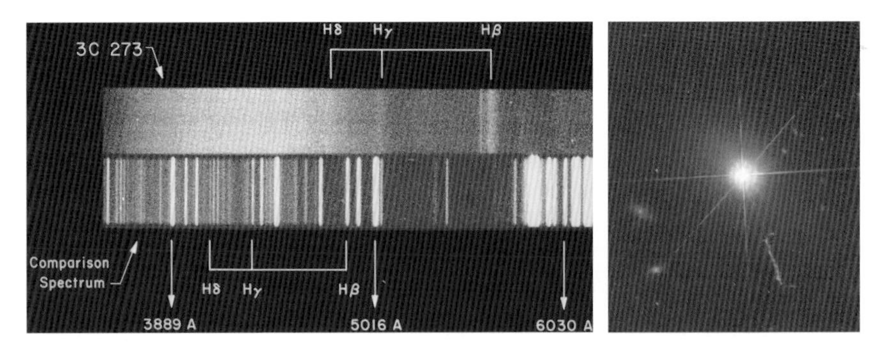

Figure 3.14 Left: the original spectrum of 3C273 obtained by Maarten Schmidt in 1962. (Image courtesy of Maarten Schmidt; original negative image converted to a positive by the author.) Right: an HST optical image 3C273. Note the star-like appearance of the main object and the single jet. (Image courtesy of NASA/STScI and the late John Bahcall.)

spectrum of the main object (Fig. 3.14). Like that of 3C48 the spectrum consisted mainly of strong broad emission lines that did not match those in any other star – including the ones in 3C48. However after much effort, Schmidt thought he recognized a pattern similar to that of the normal hydrogen spectrum, which is a series of lines whose separations reduce regularly towards the shorter wavelengths. However the series was a long way from where it should be – the second line in the series was in the yellow part of the spectrum instead of being blue–green. However unlikely it must have seemed, Schmidt assumed that it *was* the hydrogen series and worked out that it had been redshifted by 15.8%. His radical assumption quickly checked out – applying the same red-shift to the other lines in the spectrum enabled them to be identified with those from known elements as well. There could now be little doubt; 3C273 had a redshift of 0.158 and if this was due to its speed, then it was traveling at 47,000 kilometers per second away from us. If the velocity was also cosmological (but see Box 3.7) then, using a value for the Hubble constant of 71 km/s per Mpc (Sect. 1.3.1), its distance was some 2,200 Mly (660 Mpc).

In 1963 2,200 Mly was far further away than any other known object except for 3C295, though we now know that 3C273 is actually the closest quasar to us (some QSOs and BL Lacs are closer – see below and Sect. 3.2.5). Yet the object was of magnitude 13^m, visible directly through telescopes of 0.2 meters or larger. If the object was truly at this distance, its luminosity must be incredible – magnitude -26^m or 50 times brighter than the Milky Way Galaxy. 3C273 did not hold its distance record for long however, on learning of Schmidt's interpretation, Greenstein realized that the discarded redshift of 0.37 for 3C48 was valid after all. This translates to a recessional velocity of 110,000 km/s, over a third of the speed of light. Such a speed then suggests a distance of 5,200 Mly (1,600 Mpc) and an absolute magnitude of -25^m, which is 40% of the brightness of 3C273 but still many times brighter than an ordinary galaxy[16].

Even on optical images, both 3C273 and 3C48 were unresolved and appeared star-like. Since they were also radio sources, they and other similar objects rapidly came to be labeled as "Quasi-Stellar Radio Sources", which Hong-Yee Chiu then contracted down to "Quasar". Before long many more examples of quasars were turned up via searches through archives and via new surveys that looked for bluish star-like sources – though whether the object was a bluish star or a quasar then had to be determined by obtaining its spectrum. Some of these new sources were strong radio emitters, but 90% of the objects being found were radio-quiet even though they had the characteristic emission-line spectra of quasars. This second class of sources was named "Quasi-Stellar Objects" and abbreviated to QSOs. The terms Radio-Quiet Quasar (RQQ) and Radio-Loud Quasars (RLQ) may also be encountered in the literature as synonyms for QSO and quasar respectively.

---

[16]These luminosities are based upon the assumption of isotropic emission by the quasars. However, as discussed in Sect. 3.2.1 and Chap. 5, it is likely that our line of sight is close to the track of a high-velocity jet of material emitted towards us by the AGN. The radiation from the jet is likely to be increased by relativistic beaming and boosting (Box 4.1) and so the emission from the quasar is not isotropic but is enhanced in our direction. The true luminosities are therefore likely to be lower than the values quoted in the main text.

Here we must digress slightly to mention a common source of confusion within this topic. The terms "quasar" and "QSO" are widely used as synonyms for each other by some authors and "quasar" is equally widely used by others as a name for the group of objects that includes both quasars and QSOs. Quasars are also sometimes labeled as one of the types of "core-dominated radio galaxies" (Sect. 3.2.6). Since QSOs and quasars have distinctly different physical properties and may have different internal structures, in this book we will use the terms as they were first defined – quasar for the radio-loud objects and QSO for the radio-quiet objects. Although it is slightly cumbersome, for clarity, when referring the overall group comprising both types of object, the phrase "QSO and quasar" will be used. Another possible source of confusion lies in the so-called "micro-quasars" of which SS 433 is the best-known example. These are star-sized close binary systems (and should therefore more appropriately be called "nanoquasars") in which a normal star, white dwarf or neutron star orbits a black hole of a few solar masses (or possibly a neutron star in some cases). Although mass transfer to the black hole/neutron star occurs, leading to the formation of an accretion disk, and perhaps to the production of relativistic jets, they are otherwise completely unrelated to quasars or any other type of AGN.

Once astronomers knew what to look for, the number of recognized QSOs and quasars, as well as other types of AGN increased rapidly (Fig. 3.15). The AAT 2dF and the SDSS surveys (Box 3.2) have recently added greatly to the numbers, so that at the time of writing (2006) nearly 85,000 QSOs and quasars have been catalogued, of which about 90% are QSOs and the rest are quasars. The closest QSO that is currently known is in Pisces and is called QSO B0050+124 (it is also identified as I Zw 1 from Fritz Zwicky's catalogue), though with an absolute magnitude of -23^{m} it is on the Seyfert/QSO borderline (see later discussion) and is thus sometimes classed as a bright Seyfert 1 instead. It has a redshift of 0.0609, corresponding to a recession velocity of 18,300 km/s and a distance of 850 Mly (260 Mpc). As already noted, 3C273 is the closest quasar. However the redshifts of QSOs and quasars are generally very much larger than those of B0050+124 and 3C273 and

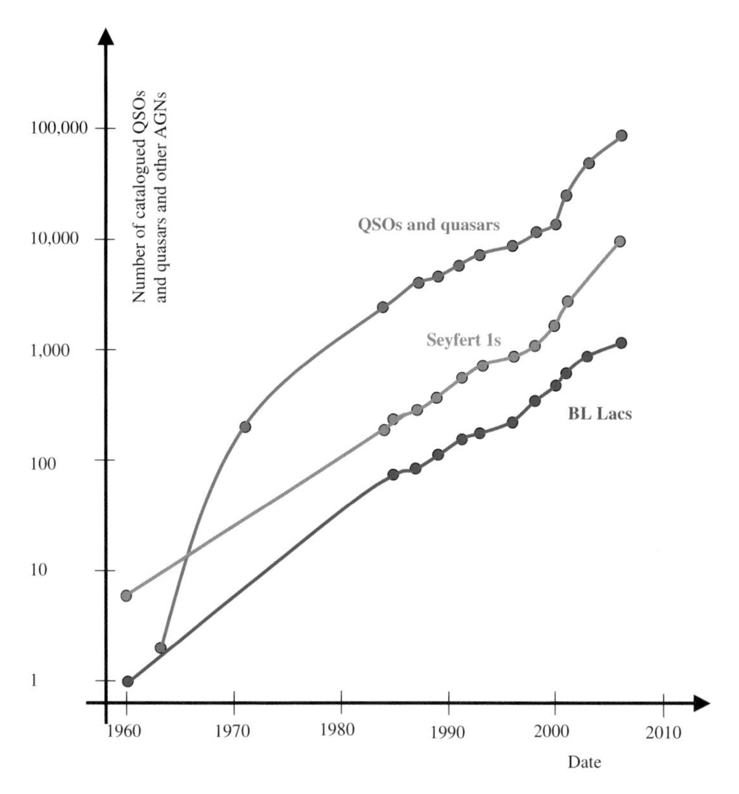

Figure 3.15 The increase in the number of catalogued QSOs and quasars, Seyfert 1s and BL Lacs with time.[17]

there is a peak in the numbers of objects for redshifts between two and three. Such redshifts translate to recessional velocities of 240,000–270,000 km/s and distances of 11,000–12,000 Mly (3,400 to 3,700 Mpc). The peak in QSO and quasar numbers is so pronounced – they were 1,000–10,000 times more numerous than they are today – that the time when the universe was around two to three aeons old is sometimes referred to the quasar era. Above redshifts of about three the numbers

[17]The author is grateful to Dr Mira Véron-Cetty of the Observatoire de Haute Provence for supplying this information in advance of the publication of the 12th edition of the *Catalogue of Quasars and Active Galactic Nuclei* (see Appendix 1). Other data are from the 11th edition of the *Catalogue of Quasars and Active Galactic Nuclei*, M.-P. Véron-Cetty and P. Véron, 2003.

of QSOs and quasars drop rapidly and they are almost non-existent beyond a redshift of five (13,200 Mly, 4,000 Mpc). The SDSS has though, thrown up a few examples (out of 85,000) with $z > 5$ and a couple with $z > 6$, including the current AGN record holder at $z = 6.4$ (13,500 Mly, 4,100 Mpc). QSOs and quasars held the record for the largest known redshifts for many years, but their decline in numbers above $z = 5$ means that the highest known redshift today ($z = 6.68$) is for a gravitationally lensed starburst galaxy[18]. The decline in QSO and quasar numbers beyond a redshift of five could be due to dust clouds within their host galaxies (see below) obscuring the more distant examples, but a genuine reduction in their numbers seems more likely. One possibility is that QSOs and quasars evolve from ULIRGs so that prior to $z = 5$, they had yet to become apparent to external observers (Sect. 2.3 and Chap. 6, also see type-2 QSOs below).

Archive images of QSOs and quasars soon after their discovery together with subsequent monitoring quickly showed that for many of them their optical brightnesses varied significantly on timescales as short as a few months to a year. This constrained the maximum size of the emitting region (Box 3.4) to about one light year or less. A small group of quasars brighten by large amounts in a few days before fading again more slowly. Called the Optically Violently Variable quasars (OVVs), they are considered together with BL Lac objects in Sect. 3.2.5. In the last few years an even tighter limit to the sizes of at least some QSO and quasar emission regions has been imposed by the observation of twinkling in gravitationally lensed (Box 2.4) images of the objects. The twinkling occurs as individual stars within the lensing galaxy move and change the effect of the lens by small amounts. It is observed to occur for the background continuous component of the QSO's or quasar's spectrum, but not for its emission lines. The continuum emission region should thus be comparable with the solar system

[18]There has recently been a claimed observation of a redshift of ten. However the data analysis that produced this result has been criticized by other workers and it may not be real.

in size (~1 light day) while the regions producing the emission lines must be considerably larger (in the same way stars twinkle but not the angularly larger planets due to the effects of the Earth's atmosphere). Rapid variability at x-ray wavelengths also constrains the sizes of many of the x-ray-producing regions to be less than a few light days – and one quasar, PKS 0558–504, has even brightened by two-thirds in just three minutes. Quasars' radio emissions are generally non-variable though at least one, PKS0404–385, has brightened and then faded again in under an hour. Generally the amplitude of the variations increases towards shorter wavelengths and decreases as the QSOs and quasars get brighter.

As we have seen, interpreting the redshift as cosmological in origin leads to enormous distances and hence to extremely high luminosities for the QSOs and quasars. Initially it seemed to some astronomers improbable that a hundred trillion or more times the Sun's energy could be generated within a region as small as one light year (never mind one light day). However if the redshifts of QSOs and quasars were not cosmological (i.e. did not arise from the expansion of the universe but were due to their actual intrinsic velocities through space) then they could be much closer to us and their luminosities would then be correspondingly reduced to more "reasonable" levels. The local hypothesis for QSOs and quasars has now been abandoned for a number of reasons. Firstly because no comparable blushifted objects have ever been found – and they would be expected to occur if, say, the QSOs and quasars originated as debris from explosions in nearby galaxies. Secondly about 0.1% of QSOs and quasars are undergoing significant gravitational lensing by foreground galaxies. In these cases at least, the QSOs or quasars must be at considerably greater distances from us than the lensing galaxies, which themselves, when observable, have redshifts consistent with their being at the appropriate cosmological distances. Then for some QSOs and quasars the host galaxy can be detected sufficiently well for its spectrum to be obtained. The redshift for the host galaxy in such cases agrees with that for its QSO or quasar. Finally the unified model for AGNs (Chap. 5) via interactions with super-massive black holes *can* generate the energy needed for QSOs and quasars within the required size limitations.

Another early proposal was that the redshift arose from neither the intrinsic velocities nor from the expansion of the universe but in some new fashion such as changes in some of the fundamental constants of physics with time or through the effects of gravity. These ideas have also now been discarded. A rather more subtle argument that QSO and quasar redshifts are not all cosmological in origin is presented by Halton Arp and his supporters and is discussed in Box 3.7.

Box 3.7 Halton Arp and Non-Cosmological Redshifts

Most astronomers today accept without question that the redshifts observed for QSOs and quasars arise from the expansion of the universe (i.e. that they are *cosmological* in origin). However widespread acceptance of a theory does not automatically make it correct – Ptolemy's Earth centered and incorrect model for the solar system was, after all, regarded in the West as completely authoritative for well over a millennium, and when some did suggest that the Earth might be moving around the Sun, the church caused them to be imprisoned (Galileo) or burnt at the stake (Giordano Bruno) for their temerity. Scientific mavericks no longer face quite such severe physical penalties, but their ideas often still receive equally short shrift. Halton Christian Arp is one such rebel who has long tried to make the case for the redshifts of QSOs and quasars (and other AGNs) originating in some non-cosmological manner.

Arp's principal evidence for his ideas is based upon observations of normal low-redshift galaxies that are close to high-redshift AGNs in the sky. Such alignments can of course just be due to chance – the nearby galaxy and the distant AGN simply happening to lie in almost the same direction as seen from the Earth. In a similar fashion binary stars are physically connected to each other while double stars are flukes whose components are actually at very different distances away from us. Over two centuries ago Sir William Herschel realized that there were too many such pairs of stars to be seen in the sky for them all to be doubles, and that most must be gravitationally coupled binary systems. Arp likewise finds that there are too many close pairings between galaxies and AGNs to be due purely to chance. In some pairs however Arp also finds what appear to be physical links between the galaxy and the AGN, thus requiring them to be physically close together in space. For example some images, although not all, of the spiral galaxy NGC 4319 ($z = 0.006$) and the AGN Markarian 205 ($z = 0.70$) show them apparently linked by a luminous bridge of material. If this is indeed the case then clearly both cannot be at their required cosmological distances – and since NGC 4319 has an appearance that is consistent with its being at its appropriate cosmological distance, it would seem that it is the AGN that is out of position. Arp has found several other comparable linked pairs and cases of double or multiple QSOs or quasars close to galaxies. Most, if not all, of the latter though result from gravitational lensing (Box 2.4). Perfect alignment between the Earth, lensing and distant galaxies results in Einstein rings while a slight misalignment produces partial rings or arcs or multiple point source images. The apparently multiple QSOs or quasars have identical redshifts, which by Arp's variable mass hypothesis (see below) would result only if they were produced at the same time, but which arises automatically from gravitational lensing since the various images are then all derived from the same original object. Additional evidence for the gravitational lens interpretation of multiple QSOs comes from the observation of identical fluctuations in their brightnesses. For example Q0957+561A and B are 17^{m} twin QSOs in Ursa Major both of which have a redshift of 1.390. In the early 1990s both underwent similar changes in brightness with a time delay between the changes of 1.48 years arising from the slightly different paths from the QSO to the Earth through the gravitational lens.

Some suggested pairings though are at angular separations too great to be explained by gravitational lensing. For example the quasar

3C273 and the radio galaxy M87 (3C274) are aligned with and on opposite sides of the elliptical galaxy M 49 with a separation of about 12°. Arp gives the probability of such an alignment being due to chance as one in a million and argues that 3C273 ($z = 0.158$) must therefore be associated with the Virgo cluster of galaxies ($z \approx 0.003$) to which the other two galaxies belong. Such a statistical argument sounds very persuasive, especially when Arp has found a number of other such alignments such as 3C48, M 33 and PKS0123+25 and NGC 4258 with two quasars that are strong x-ray sources. However although a chance of one in a million seems unlikely, when there are several tens of thousands of the various types of AGN and tens of millions of other types of galaxy, the number of pairs of objects, even if their separation is limited to a maximum of 12°, is not millions, nor billions but trillions. If lines are then drawn between such pairs of objects, the probability of a third "significant" object lying close to one of those trillions of lines becomes extremely high, even when the chance for any individual pair is very low. Thus the occasional triple, quadruple or perhaps even higher multiple alignment of extragalactic objects *is* to be expected purely through chance. (The same phenomenon underlies the alignments of churches, historical sites, hills, tracks, etc., which are fallaciously used to argue for the existence of Ley lines on the Earth).

More seriously for the cosmological interpretation of redshifts, there is a recently discovered example of a QSO ($z = 2.11$) apparently being silhouetted against or within a lower redshift Seyfert 2 galaxy. The galaxy is NGC 7319 which is a member of the Stephan's quintet group of galaxies and has $z = 0.023$. The redshifts of both the galaxy and QSO were obtained by a team led by Margaret Burbidge using the 10-m Keck telescope. The QSO in this case does not appear to masked by gas and dust clouds within the galaxy but does show signs of apparently interacting with the interstellar gas. It is conceivable that the QSO lies behind the galaxy and is simply shining through a low-density region, but if this is not the case, then it seems a strong candidate to be an AGN that is not at its cosmological distance and critics of Arp's ideas have some serious explaining to do.

Of course it is not only Arp's critics who may not have all the answers – Arp's ideas also have serious difficulties. In particular, if the redshifts are not cosmological then how do they arise? As remarked in the main text if QSOs, etc. are expelled somehow from galaxies, then some should be approaching us and so have blueshifted spectra – and this is not the case. Arp and others have therefore proposed the variable mass hypothesis in which QSOs are born in and ejected from the nuclei of their parent galaxies whilst containing particles that have smaller masses (perhaps zero masses) than the sub-atomic particles observed on the Earth. The particle masses then increase with time and lead to the observed redshifts. The theory is based upon Mach's principle that inertia is due to the influence of all the matter in the universe. An electron (say) created a year ago will only be "aware" of the matter within a volume one light year in radius. After 10 years, it will be "aware" of the matter in a volume 10 light years in radius, and so on. Thus the amount of matter influencing the electron's inertia increases with time and so also does its mass. The variable mass hypothesis is pretty weird, but then ideas in cosmology such as inflation and dark energy that result from the conventional interpretation of redshifts (amongst other things) are pretty weird as well, so that does not automatically rule it out.

Another observation that is difficult to explain via Arp's model is the presence of the Lyman-α forest (see main text) in the spectra of QSOs and quasars. The absorption lines in the forest are produced by gas clouds lying between the QSO and ourselves which all have different and smaller redshifts than the QSO. This is automatically the result if the gas clouds' and QSO's redshifts are all cosmological, but the variable mass hypothesis would require that the clouds all be of different ages and all be older than the QSO – which is highly unlikely given the huge numbers of lines in Lyman-α forests and the many such forests that have been examined. Finally the redshifts of host galaxies, when these can be measured, are found to be reasonable matches to those of their QSOs or quasars, as are the redshifts of those clusters of galaxies that contain a QSO or quasar – strong support in these cases for the QSO's or quasar's redshift being cosmological in origin.

Unfortunately debate about whether redshifts are cosmological or not has become absurd – supporters of the mainstream interpretation largely just ignore Arp's evidence and ideas, while his supporters defend his ideas fanatically, rejecting any contrary evidence that may emerge. Arp is making a case that needs better answers than it is receiving but it seems likely that time will eventually prove that he is one rebel who does not have a cause. Readers interested in pursuing this subject further are referred to Arp's book *Seeing Red: Redshifts, Cosmology and Academic Science* published by Aperion, 1998.

A consequence of large redshifts is that the light that we receive within the visible part of the spectrum of a QSO or quasar was originally emitted at much shorter wavelengths. The term "rest wavelength" is used to indicate when it is the originally emitted wavelength that is being discussed and "observed wavelength" to indicate what we actually collect in our telescopes. A strong emission line from hydrogen, which has a rest wavelength of 121.5 nm and is known as Lyman-α, is moved to an observed wavelength of 350 nm (where the Earth's atmosphere starts to become transparent) by a redshift of 1.9. Since the peak in QSO and quasar numbers lies at higher redshifts than this, many of them have been discovered through searches that employ low-dispersion spectroscopes to hunt for Lyman-α in optical spectra. The range of redshifts also means that the characteristics of QSO and quasar spectra from well into the ultraviolet through to the near infrared were known before it was possible to observe their ultraviolet regions directly. The emission lines originally emitted by QSOs and quasars at visual wavelengths are broad and strong and closely resemble those in Seyfert

1 spectra (Fig. 3.8). While in the ultraviolet, Lyman-α is usually the strongest line followed by Mg II 279.8, C III] 190.9, C IV 154.9 and O VI 103.4. There is also a blend at 140 nm due to O IV] 140.2 and Si IV 139.7 while Lyman-α is blended with N V 124.0. Somewhat puzzlingly, the strength of the C IV 154.9 is found to become weaker relative to the underlying continuum intensity as the overall luminosity of the AGN increases. Called the Baldwin effect, the reasons why the line strength should behave this way are unclear, however it does enable the luminosity of the AGN to be estimated independently of the redshift. A similar effect also occurs for some of the x-ray emission lines due to iron.

Absorption lines in QSO and quasar spectra divide into the intrinsic and extrinsic types. The intrinsic lines are produced by material within the QSOs or quasars or by other material that is physically associated with them and sub-divide into the broad and narrow types. About one in ten QSOs have spectra containing broad absorption lines – and hence are sometimes called BALQSOs. The broad lines have widths corresponding to Doppler broadening velocities up to 10,000 km/s and may be blueshifted[19] with respect to the emission lines by up to several tens of thousands of kilometers per second, suggesting that they originate in rapidly out-flowing material within the QSO. A few quasars have recently been found with broad absorption lines although their blueshifts are lower than those in BALQSOs. Around half of QSOs and quasars have narrow intrinsic absorption lines in their spectra that have widths corresponding to a few hundreds of kilometers per second and these usually have redshifts close to those of the emission lines.

The extrinsic absorption lines are all very narrow and are due to absorption by gas clouds lying along the line of sight. The broader of the extrinsic absorption lines, which are still a lot narrower than the broad lines in BALQSOs, are thought to be due to the gas in galactic

[19]They are not blueshifted when seen from the Earth – just less redshifted than the emission lines.

haloes whilst the other lines arise from absorption within gas clouds in intergalactic space. The gas clouds are unobservable apart from their production of these absorption lines, so the lines are mostly studied for the information they reveal about conditions between the galaxies. Since the gas clouds must clearly be closer to us than the QSO or quasar, their cosmological redshifts are smaller and the extrinsic lines occur at shorter wavelengths than their intrinsic equivalents. This is particularly noticeable for the Lyman-α line. In some QSO and quasar spectra the region shortward of the intrinsic Lyman-α emission line is packed with dozens, sometimes hundreds, of narrow extrinsic Lyman-α absorption lines. The number of such lines has led to this region sometimes being labeled the "Lyman-α forest". Each line arises within a separate gas cloud and so there must be hundreds of intergalactic or galactic halo clouds along the lines of sight to the more distant QSOs and quasars. Data from the Lyman-α forest suggest that individual clouds have masses of 10–100 megasuns and are around 30,000–40,000 ly (10 kpc) across.

The continuous spectra of QSOs and quasars are generally similar to those of Seyfert 1s with the intensity reducing from a peak in the infrared through to the optical. Some then exhibit "big blue bumps" similar to those found in some Seyferts arising from second peaks for their emission in the ultraviolet. Many QSOs and quasars also emit strongly in the x-ray region, particularly within the softer (longer wavelength) part. To a fairly rough approximation QSOs and quasars have a flat spectrum (i.e. constant luminosity) from the infrared to the x-ray regions, in contrast with normal galaxies where the thermal emission from stars results in a continuum that peaks strongly in the visible and near ultraviolet regions. In recent years the Compton gamma-ray observatory and other gamma-ray telescopes including ground-based Čerenkov instruments[20], have found a hundred or so QSOs and quasars

[20]When a high-energy gamma ray encounters the top of the Earth's atmosphere it interacts with nuclei in the atmosphere to produce a shower of high-energy sub-atomic particles. High in the atmosphere the speed of light is only

emitting powerfully within the very high-energy gamma-ray region. In most cases the photons have energies up to a few hundred MeV, but a few cases have been found with TeV emissions.

Quasars clearly differ from Seyferts in having very strong emission at radio wavelengths and this may extend far enough into the infrared to swamp the normal emission peak there. Typically the radio brightness of a quasar is 100 times that of a QSO. The radio emission seems to be associated with the possession by the quasars of a jet of relativistic electrons (Chap. 4). Sometimes, as for 3C273, the radio emission from the jet can be picked up separately from that of the main body of the quasar. Very long baseline interferometry (VLBI) shows that quasar radio emission regions range from milliarcseconds to tens of arcseconds in size – corresponding to physical sizes ranging from 100 ly (30 pc) to a million light years (300 kpc) for quasars at a distance of 10,000 Mly (3,300 Mpc). As well as radio emission from the central object, some quasars such as 3C47 (Fig. 3.16), identified in 1964, have extended double-lobed radio emission regions like those of the FR 2 radio galaxies (Sect. 3.2.6).

Levels of polarization in the radiation from quasars are low (3% or less) and even lower for QSOs. This is in part, though, because quasars with high levels of polarization are classed as Blazars (Sect. 3.2.5) and so considered to be a separate type of object. However in some quasars high levels of polarization arising from a synchrotron radiation source (Box 3.1) could be being swamped by stronger unpolarized emissions coming from thermal or free–free radiation sources (Boxes 2.5 and 3.1), so leading to a low overall level of polarization.

---

somewhere in the region of ~0.000,01% to ~0.001% (~30 to ~3,000 m/s) less than the speed light in a vacuum. Nonetheless the particles produced by the gamma ray are moving faster than the speed of light in the atmosphere. This results in a shock wave analogous to the sonic boom that is created when the speed of sound in the atmosphere is broken. In this case the "boom" takes the form of a flash of light that can be detected from the ground and so the presence of the gamma ray inferred. The flash of light is called Čerenkov radiation after the Russian physicist Pavel Čerenkov who first studied it.

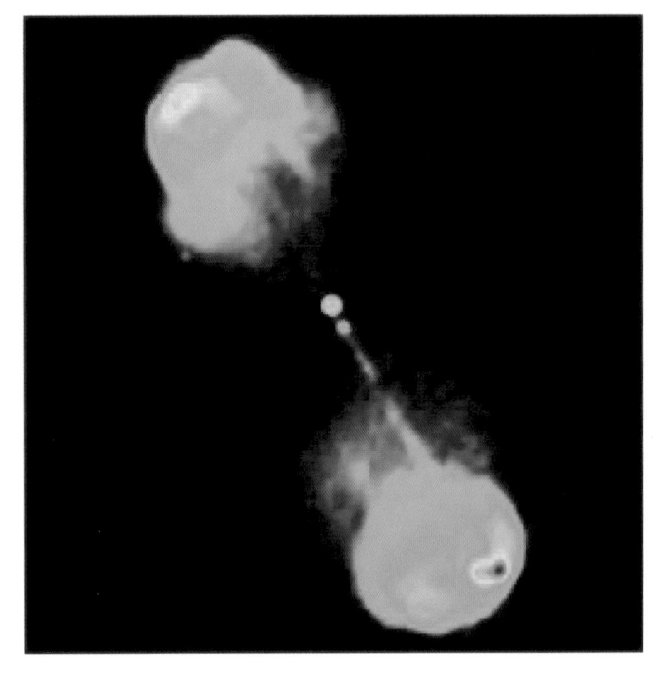

Figure 3.16 A radio "photograph" of quasar 3C47 showing the extended radio structures that cause it to be classified also as an FR 2 double-lobed radio source. (Image courtesy of Patrick Leahy.)

This may well be the case for the first known quasar, 3C273, which perhaps should thus be classed as a high-polarization quasar (HPQ) type of Blazar (Sect. 3.2.5) rather than a normal quasar.

QSOs and quasars were initially observed optically as point sources, hence the "quasi-stellar" part of their name. That is still generally the case today, although 3C48 was seen from the start to be centered within a larger faint nebulous emission region that turned out to be an interacting galaxy and 3C273 was quickly found to be within an elliptical galaxy. However it is only in the last decade or so that the high resolutions of the Hubble space telescope and of 8-m class ground-based telescopes using active atmospheric compensation have enabled a few tens or so more QSOs and quasars to be found which possess surrounding "fuzzes". Although the observations are difficult due to the glare from the main object, the "fuzzes" turn out to be host galaxies that contain the QSOs or quasars within themselves. The host galaxies are likely to be larger than average and can be spiral, elliptical or disturbed in form with the latter often showing signs of an interaction with another galaxy such as starburst activity. The QSOs tend to have spirals or ellipticals as hosts, while the quasars are located within elliptical or disturbed galaxies. It is widely expected that all QSOs and quasars would be found to lie within host galaxies if only good enough observations could be obtained, since just such a surrounding galaxy is needed to provide fuel for the central black hole energizing the AGN (Sect. 3.2.1 and Chap. 5).

The visual rest spectra of QSOs strongly resemble those of the nuclei of Seyfert 1 galaxies (Sect. 3.2.2). If the host galaxy containing a Seyfert 1 nucleus could not be seen, then it would be very difficult to find any other criterion to distinguish it from a QSO. Now, as just discussed, some, possibly all, QSOs and quasars have host galaxies, so other than QSOs being brighter than the nuclei of Seyfert 1s, there seems to be little difference between them. For this reason it is now generally accepted that there is continuity of type between Seyfert 1 galaxies and QSOs (and by extension, with quasars as well). Based upon the historical distinction that Seyferts are (relatively) faint and QSOs and quasars bright, it has become customary to use an absolute V[21] magnitude of -23^{m} to separate the classes – Seyferts being fainter than this and QSOs and quasars brighter. However this is a quite arbitrary definition and one that is not universally followed (particularly in the past) so that AGNs brighter than -23^{m} may be encountered being classed as Seyferts while others, fainter than the limit, are regarded as QSOs or quasars. A separate though less widely used criterion is that QSOs and quasars have redshifts larger than 0.1 so that objects with redshifts smaller than this are taken to be another type of AGN. There remains however one puzzle in interpreting Seyfert 1s as faint QSOs –

[21]The "visual" magnitude within the widely used UBV photometric system for measuring the brightnesses of stars, etc. It actually covers just the yellow part of the spectrum.

Seyfert 1s are predominantly found within spiral hosts, while QSOs lie within both spirals and ellipticals. The reason for this difference is presently unclear and it may turn out to have great significance, or it could simply be the result of some bias within the relatively few observations that there are of QSO hosts.

As we have seen in earlier sections, faint Seyferts may merge with bright LINERS and faint LINERS be almost indistinguishable from normal galaxies, we thus apparently have a sequence:

Normal galaxy – LINER – Seyfert galaxy – QSO and quasar

along which there is little real difference except for the level of activity (i.e. luminosity) of the AGN. Although physical validity of such a linked sequence is not yet proven, as we shall see (Sect. 3.2.1 and Chap. 5), a plausible explanation for the sequence lies in an increasing mass for the central massive black hole and an increasing availability of fuel. The possibility that some Seyfert galaxies are the last stages of QSOs and quasars (i.e. "dying" QSOs and quasars) cannot be ruled out, although there seem to be too many Seyferts for them all to have evolved from QSOs and quasars. Also, where they can be determined, the masses of the central black holes tend to be lower in Seyfert galaxies than they are in QSOs and quasars, suggesting that the difference is not simply that of age.

The emission lines in QSOs' and quasars' spectra are broad, making them type-1 objects (Sect. 3.2.2). The question of whether or not they have type-2 equivalents is still open although the answer "yes" now seems more likely than "no". The difference between Seyferts of types 1 and 2 is that the BLR is hidden, perhaps by thick gas clouds, in the latter. A type-2 QSO or quasar is thus also called one that has a hidden or buried BLR.

The numbers of buried QSOs and quasars, if they exist at all, seem to be small compared with the numbers of the type-1 objects, suggesting that their lifetimes are very short (though evidence to the contrary may come from x-ray observations; see below). The reason for this may simply be that QSOs and quasars are sufficiently energetic that they evaporate or blow away their dusty shrouds within a short time, then becoming visible as "normal" type-1 objects. Type-2 Seyferts would thus be accounted for by having AGNs that were not sufficiently energetic to do the same thing. As with the Seyferts though, it may be that the differences between type-1 and type-2 QSOs and quasars arise from their orientation to the line of sight.

If the lives of buried QSOs and quasars are short and they evolve into their normal equivalents, then this suggests that we should look for them amongst objects at distances where the look-back time shows them to us as they were during the first few aeons of the universe's existence. The ULIRG (Chap. 2) FSC 10214+4724 easily meets this requirement with a distance of 11,500 Mly (3,500 Mpc). Although gravitational lensing (Box 2.4) may have enhanced its apparent brightness, it is still amongst the brightest objects in the universe and on these grounds alone has long been proposed as a type-2 QSO. Strong support for that has come from the detection of broad emission lines in the polarized component of its spectrum. Like the similar observations for some Seyfert 2s (Sect. 3.2.2) this is interpreted as radiation from the central BLR of the QSO or quasar that has been scattered towards us. As we saw in Chap. 2 many more ULIRGs are suspected of being buried QSOs and quasars – perhaps up to two-thirds of them if the spectroscopic evidence from ESO's VLT is confirmed.

Buried QSOs and quasars should be directly observable at hard x-ray wavelengths since these photons can penetrate the dust clouds. It should also be possible to separate the QSOs and quasars from other x-ray sources since their soft x-ray emission will be absorbed. They should thus have a characteristic signature of being a bright hard x-ray source and a faint soft x-ray source. The Chandra spacecraft has recently observed two such sources. One is found at the center of a spiral galaxy in Pegasus, the other was located during the Chandra deep field south exposure and is in Fornax at a distance of 12,000 Mly (3,600 Mpc). The Spitzer infrared spacecraft too has found a number of possible type-2 QSO and quasar candidates from their infrared emissions.

Spitzer has also observed emission at 10- and 18-micron wavelengths in some type-1 QSOs and quasars coming from hot dust particles – perhaps the remnants of an earlier shroud. Additionally there is indirect evidence that type-2 QSOs and quasars may be quite numerous. Background x-ray radiation is thought to originate as the x-rays emitted by AGNs, with QSOs and quasars making the largest contribution. However the numbers of type-1 QSOs and quasars seem to be too low to produce the observed intensity of the background. It has been suggested that x-rays from type-2 QSOs and quasars make up the deficit. If this is the case, then the type-2 objects must be far more numerous than the type 1s.

One final feature of quasars that needs mentioning is their possession of jets of relativistic electrons. Often only one jet is detectable though two are to be expected moving in opposite directions. Relativistic effects however will mean that the jet moving away from us is much fainter than the approaching jet and so less easy to see. This is the case with the first quasar, 3C273, and its jet can be clearly seen separately from the main body of the object (Fig. 3.14). Jets occur in many other types of AGNs and so are dealt with as an independent topic (Chap. 4) rather than repeat the discussion several times.

3.2.5 Blazars

The term "blazar" was coined as a generic term for two types of AGNs with a number of observational features in common, but which in fact possess different natures. The AGNs involved are the BL Lac objects and the High-Polarization Quasars (HPQs). A number of the latter are highly changeable and so the Optically Violently Variable quasars (OVVs) are sometimes identified as a sub-class of the HPQs. Here though, we shall treat HPQs and OVVs as being the same type of object. BL Lacs and HPQs are also sometimes regarded as varieties of "core-dominated radio galaxies" (Sect. 3.2.6).

In 1929 a star-like object in Lacerta close to its border with Cygnus was discovered to be the 101st variable in the constellation. Following the usual practice of variable star observers[22] it was named BL Lac. For nearly four decades thereafter it remained in obscurity until in 1968 it was found also to be an extremely variable and angularly small radio source. Optical spectra soon showed that it was a real eccentric – its spectrum was featureless – no emission lines and no absorption lines, i.e. a continuous spectrum (Fig. 1.3) just as though it were a (very) hot solid. The intensity of the continuous emission increased markedly towards the red and into the infrared parts of the spectrum and it was all strongly linearly polarized.

Higher quality spectra obtained during BL Lac's fainter phases eventually did show the presence of very faint emission lines and enabled its redshift to be found to have a value of 0.069. This is a small redshift compared with some of those that we have just encountered for QSOs and quasars (Sect. 3.2.4), but nonetheless, if interpreted cosmologically, it corresponds to a distance of 950 Mly (300 Mpc). The escape velocities from galaxies are a few hundreds of kilometers per second, so if BL Lac were to be a relatively nearby object, its velocity of nearly 21,000 km/s would soon ensure that it left any parent galaxy far behind. Clearly BL Lac was not a peculiar star within our own Galaxy.

Over the next decade or so a few more objects with properties similar to those of BL Lac were discovered. Mostly these were found from their radio spectra, which showed them to be compact sources with flat and variable radio emissions (see below). One further case of an object that had previously been thought to be a variable star turned up though. A 15^{m} variable towards the southern edge of Libra, named AP Lib, actually turned out to be a BL Lac object with a redshift

[22]The naming of variable stars is unnecessarily complex, however it seems unlikely that a more rational system will now be devised, so readers interested in finding out how it works are referred to, amongst other sources, the author's book *Stars, Nebulae and the Interstellar Medium* (page 128), published by Adam Hilger, 1987.

of 0.049 and so at a distance of 680 Mly (210 Mpc). However they were discovered though, these new objects were all given the generic name of BL Lac objects (or Lacertids, although the latter name is not widely used) and added to the growing collection of AGNs. Further study showed that their optical and infrared brightness changes could be by more than a factor of 10 and occur in just a few days. At shorter wavelengths the changes could happen in just a few hours. The degree of linear polarization of their emissions also varied rapidly and might well be up to 40% in value. The intensity of their spectra continued to increase with increasing wavelength up to a maximum in the far infrared or microwave region. At radio wavelengths the intensity then became almost constant, i.e. a "flat" spectrum. In some BL Lacs more extended radio emission is also found and this resembles that from the FR 2 class of radio galaxies (Sect. 3.2.6). A few of the BL Lacs, including BL Lac itself, were found to be surrounded by faint fuzzes, and spectra of the fuzzes revealed them to be mostly elliptical galaxies with redshifts similar to those of their AGN components (when these could be measured). The possibility that BL Lacs might be (relatively) local was thus soon discounted. The lowest redshift BL Lacs are currently the 13^m objects Markarian 421 in Ursa Major ($z = 0.031$) and Markarian 501 in Hercules ($z = 0.034$, see Fig. 3.17), corresponding to cosmological distances of 430 Mly (130 Mpc) and 470 Mly (140 Mpc) respectively.

The early discoveries of BL Lac objects were based upon their radio emissions, and so they were all radio-loud. Once spacecraft-based x-ray observations became available, the x-ray intensities of these BL Lacs were found to continue the trend of decreasing intensity with decreasing wavelength found in their optical and infrared spectra. However the x-ray observations also revealed a new type of BL Lac that was bright in the x-ray region, but which had only about 1% of the radio intensity of the "normal" BL Lacs. This new class of BL Lacs though still had a radio intensity higher than that of radio-quiet objects such as QSOs. They are thus sometimes called radio weak BL Lacs, but are more commonly known as x-ray loud BL Lacs (XBLs) or some-

Figure 3.17 The second closest BL Lac object to us, Markarian 501 (the slightly fuzzy object just below center). For several months in 1997 this radio-loud BL Lac object (RBL) was simultaneously an x-ray loud (or at least gamma-ray loud) BL Lac (XBL) as well. (Image courtesy of Leslie F. Brown, Jenna Beam, Olin Observatory, Connecticut college.)

times as High-frequency BL Lacs (HBLs). The original group of BL Lacs thus now become the radio-loud BL Lacs (RBLs) or Low-frequency BL Lacs (LBLs). The emissions from XBLs peak towards the short wavelength end of the ultraviolet region, with a second peak in the far infrared. In the optical region the polarization levels tend to be lower for XBLs than RBLs and the XBLs tend to be less strongly variable. XBLs overall are more numerous but fainter than RBLs and thus tend to be found in relatively close proximity to the Milky Way Galaxy.

The distinction between RBLs and XBLs may however be more apparent than real. In 1997 Markarian 501, the second closest BL Lac to us and a strong radio source, suddenly flared in the gamma-ray region until its apparent gamma-ray luminosity was 10 times brighter than that of the Crab nebula (the brightest non-variable gamma-ray source in the sky). In absolute terms Mrk 501 was then 5,000 million times brighter than the Crab. Furthermore the outburst lasted for several months and gamma rays were observed with energies up to 20 TeV. BL Lac itself has also at times been a strong gamma-ray emitter, as have several other RBLs.

Milliarcsecond-resolution radio observations by VLBIs of the cores of some BL Lacs, including BL Lac itself, have picked up bullets of material apparently moving at several times the speed of light. These super-luminal motions are considered in detail in Chap. 4. Here we just note that the speed limit of the speed of light that arises from special relativity is not in fact being broken in these BL Lacs. The super-luminal velocity is an illusion that results when a jet of material is moving almost directly towards the observer at speed that is close to, but less than, the speed of light.

The number of known BL Lac objects of both types now exceeds a thousand (Fig. 3.15), but that figure needs to be treated with some caution. Firstly for some BL Lacs, including BL Lac itself, the emission-line intensities can vary from zero at some times to being quite strong at others. The emission lines tend to be seen when the BL Lac object itself is faint, suggesting that they are always present but are swamped by the continuous emission when the BL Lac object is bright. Clearly, if a BL Lac object were to be observed during one of its strong emission-line phases, it would probably be classed as another type of AGN. Secondly, because the definition of something being a BL Lac object depends on *not* being able to detect emission lines, poor quality spectra (i.e. spectra with low dispersions and/or low signal-to-noise ratios) are likely to suggest that an object is a BL Lac when better spectra might show it to be another type of AGN whose emission lines were weak but not absent.

Apart from the absence of emission lines, BL Lac objects have a lot in common with QSOs and quasars. However the real clues to their natures lie in their highly polarized and continuous spectra and the super-luminal motions. Although thermal emission is continuous, it is completely unpolarized – only synchrotron emission (Box 3.1) is both continuous and polarized. Thus BL Lac's optical and longer wave emissions are almost certainly due to synchrotron radiation. Their shorter wavelength emissions are thought to be arise from the inverse Compton scattering (Box 3.1) of lower energy photons by the same high-energy electrons that are producing the synchrotron emissions. Finally the super-luminal motions show that we are looking almost directly down a relativistic jet of material, i.e. a pheasant's view of a shotgun blast, though fortunately we are sufficiently far away from the jet to escape the pheasant's fate.

Thus, given that we have seen that some quasars possess jets (and so do radio galaxies, see Sect. 3.2.6), we may picture BL Lac objects as being AGNs wherein our line of sight is almost directly along their black holes' rotation axes (Sect. 3.2.1). There is a jet blasting outwards at a speed close to that of light along the rotation axis and heading for us (there is probably also a counter-jet moving away from us on the other side of the AGN, but that is hidden). The jet is emitting continuous radiation at all wavelengths through the synchrotron and inverse Compton processes and this dominates the observed emissions. However the emissions from the jet are erratic and sometimes it fades sufficiently for the emission lines from the underlying AGN to become apparent.

The question of the luminosity of BL Lac objects has been left until we had established that their emissions probably arise from relativistic jets. Taking BL Lac's apparent visual magnitude as 14.7^{m} and its distance as 950 Mly (300 Mpc), we deduce an absolute visual magnitude of −22.7, almost bright enough to be defined as a quasar. Several other BL Lacs are significantly brighter than this and, of course, they all emit even more energy in the infrared and at longer wavelengths. Thus it would seem that BL Lacs' luminosities are comparable with those of the brightest Seyferts and with the fainter QSOs and quasars. However these calculations assume that the BL Lacs' emissions are isotropic (i.e. the same intensity in all directions), but the emission from relativistic jets is not isotropic, it is focused into narrow beams along the directions of the jets. The cause of this beaming is relativistic aberration, which is discussed in Box 4.1 Assuming the emission to be isotropic thus overestimates the luminosities of BL Lac objects by a large factor. If the material in a jet is moving towards us at 90% of the speed of light, then the emitted beam of radiation is about 30° wide, but this falls to 20° at 0.95*c* and to 10° at 0.99*c*. Remembering that there are probably two jets, this means that the beams from a single BL Lac cover about 4% of the whole sky for a jet velocity of 0.9*c*, and just 1% and 0.4% at 0.95*c* and 0.99*c* respectively. Typically therefore we see that BL Lacs' luminosities are overestimated by about a factor of 100 if their emissions are assumed to be isotropic, and so they actually become comparable with faint to typical Seyferts rather than with QSOs and quasars. The beam is also likely to have been brightened through relativistic boosting (Box 4.1), thus further exacerbating the overestimation of blazars' luminosities.

HPQs and OVVs (hereinafter just HPQs) have, as the names suggest, levels of linear polarization that can be as high as 20% and luminosities that can vary erratically by factors or 10 or more on timescales of days to weeks. They are radio-loud objects with flat radio spectra and their luminosity increases from the optical towards the infrared. They also display super-luminal motions within their cores that can, in some cases, significantly exceed the speeds reached within BL Lacs – values of nearly 30*c* having been observed. In most of these properties they clearly resemble the RBLs, but there is one major difference – HPQs' optical spectra exhibit broad and strong emission lines like those of the QSOs and quasars. Though in some HPQs the emission lines may be variable and even disappear at times

Other differences between HPQs and BL Lacs include a tendency for the HPQs to be brighter overall, so that many are observed at higher redshifts (out to $z = 5.5$) than where the BL Lacs are commonly to be found ($z < \sim 0.2$). The HPQs also tend to be the relatively more luminous of the two objects over the x-ray region. The interpretation of HPQs, like that for the BL Lacs, is via the phenomena arising through looking almost exactly down the line of an approaching relativistic jet. The differences between HPQs and BL Lacs may lie with their parent galaxies[23] – thought to be FR 2 radio galaxies for the HPQs and FR 1 radio galaxies for the BL Lacs (Sect. 3.2.6) although there may be some overlap.

The first quasar to be discovered, 3C273, possesses most of the properties of HPQs – strong and broad emission lines, radio-loud, rapid variability, super-luminal velocities, etc., but its levels of polarization are low. In the radio and microwave regions the level of linear polarization is a few per cent, but this falls to almost zero in the infrared. One possible explanation for 3C273 is that it is an HPQ, but that the synchrotron radiation is swamped most of the time within the infrared and at shorter wavelengths by a strong unpolarized continuous emission – perhaps thermal radiation coming from hot dust. Some confirmation of this model is given by the observation that when 3C273 is at its brightest, i.e. when the synchrotron radiation source is undergoing an outburst – low levels of polarization can be found in its infrared spectrum. If 3C273 is a hidden HPQ then it is possible that there are many other HPQs to be found amongst otherwise apparently "normal" quasars.

3.2.6 Double-Lobed Radio Galaxies

Many of the AGNs studied so far, such as NLRGs, BLRGs, quasars and blazars, are strong radio sources with the radio emission coming predominantly from the compact central object. They are thus often put together as a single class called the "core-dominated radio galaxies". Since we have already covered the properties of these AGNs, they will not now be considered further. In this section we are concerned with a

[23]As used in the context of blazars, "parent" does not imply something from which the BL Lac object evolved, but what the BL Lac object would look like if were to be observed well away from the direction in which the jet is moving.

second type of radio galaxy. These have one or more radio emission regions that are physically large (thousands to millions of light years across, kpc to Mpc). The radio emission regions are usually double and are commonly called the radio lobes. This type of AGN therefore goes by the name of "double-lobed radio galaxy" or "lobe-dominated radio galaxy". Some core-dominated radio galaxies, especially quasars such as 3C47 (Fig. 3.16), may also have radio emission lobes, but these are usually of lower luminosity than the central sources in such AGNs' cores. We should perhaps note at this point that all strong radio emissions from galaxies originate as synchrotron radiation.

An older sub-division of radio galaxies is into "steep spectrum" and "flat spectrum" types, corresponding roughly to the lobe-dominated and core-dominated sources respectively. The steep spectrum is just that of synchrotron radiation within the transparent regime (Fig. 3.2). Flat spectrum sources also arise from synchrotron emission, but have contributions from several different sources. The conditions (electron speeds, magnetic field strengths, etc.) vary between the sources and so their spectra peak at different wavelengths. In many cases the peak emissions are sufficiently well separated in wavelength that the combined spectrum is (very) approximately constant – the peak from one source coinciding with weak emission from another, etc., i.e. a flat overall spectrum.

In the late 1930s and early 1940s Grote Reber made the first systematic studies of radio emissions from objects in the sky using a 9.4-m paraboloidal dish operating at a wavelength of 1.9 m (160 MHz). Apart from confirming Karl Jansky's discovery of radio emission from the center of our Galaxy, he also detected the Sun and bright sources in Cassiopeia and Cygnus. The latter were simply labeled Cas A and Cyg A. Cas A we now know to be a supernova remnant within the Milky Way Galaxy. The slightly fainter Cyg A turned out to be the brightest extragalactic radio source in the sky. It was later picked up during the third Cambridge radio survey and so is also known as 3C405. With improved resolution, Cyg A was observed to comprise two extended radio sources separated by about a minute of arc with a third compact source midway between them and to which they were linked by narrow filaments (Fig. 3.18). In 1951 the compact source was identified with a 15^m giant elliptical galaxy that has a prominent central dust lane. Cyg A's redshift was measured to be 0.057, leading to a distance of 800 Mly (240 Mpc) and so to a physical separation between the two main radio sources of about 400,000 ly (120 kpc). The radio output at around 10^{39} W from Cyg A is a million times more intense than 10^{33} W of radio energy emitted by the Milky Way Galaxy (in fact the radio output from Cyg A is 100 times the *total* luminosity of the Milky Way Galaxy). Cyg A's radio spectrum is more or less flat and this is attributed to the emissions arising from a combination of synchrotron radiation sources (Box 3.1, also see below). More recently Cyg A has also been found to be an x-ray source as well.

Cyg A was just the first of many double-lobed radio galaxies to be discovered and it has turned out to be typical of them, although brighter than most. The basic properties of double-lobed radio galaxies are thus:

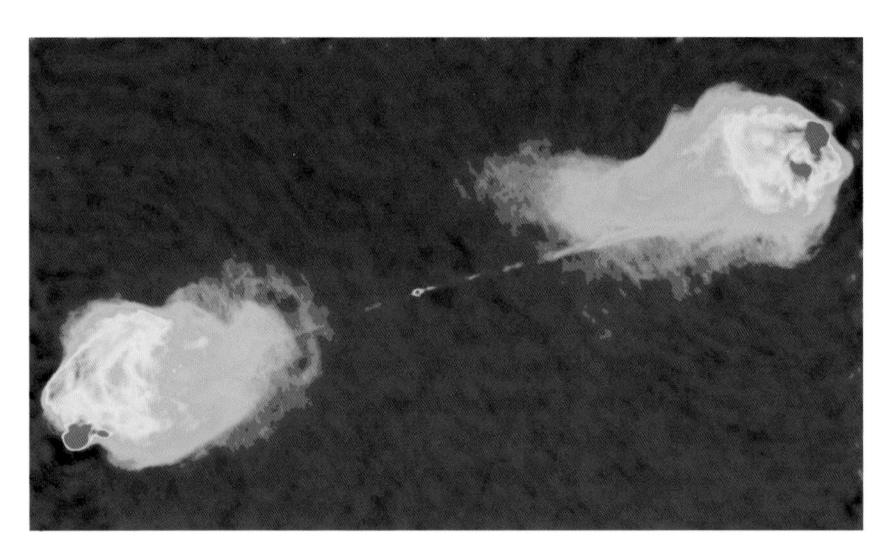

Figure 3.18 A radio "photograph" of the radio galaxy, Cyg A, showing the double lobes, central compact source and the connecting filaments. (Image courtesy of NRAO, R. Perley, C. Carilli and J. Dreher.)

(a) Two bright radio emitting lobes whose sizes may be up to several million light years (Mpc) across and which are separated by up to several million or even tens of millions of light years (Mpc). Sizes and separations around 100,000 ly (30 kpc) are more normal though.
(b) A compact and fainter radio source situated midway between the radio lobes that at optical wavelengths is seen to be a giant elliptical galaxy.
(c) Radio-emitting filaments (now usually called jets) linking the compact central source and the radio lobes. Sometimes only one jet may be detectable (as is the case for quasars).
(d) A radio spectrum characteristic of synchrotron radiation.

Cen A, the brightest radio source in Centaurus, is just such a double-lobed radio galaxy (Fig. 3.19). It is the closest double-lobed radio galaxy to us at just 11 million light years (3.5 Mpc) away. Its central radio source is identified with a 7^m giant elliptical galaxy (NGC 5128) that, like the Cyg A galaxy, is bisected by a central dust lane. Its radio lobes can be traced over 11° across the sky, though the main emission regions are separated by about 1.25°, corresponding to a physical separation of about 250,000 ly (75 kpc). Jets linking the central radio source with the lobes have been observed at both radio and x-ray wavelengths and x-ray emission has also been detected from the central core. High-resolution VLBI imaging shows the inner parts of the jets to be formed from a number of knots and that the jets' direction is changing slowly with time – perhaps due to the precession of the rotation axis of the central black hole (Sect. 3.2.1 and Chap. 5).

In 1974 Bernard Fanaroff and Julia Riley proposed dividing the double-lobed radio galaxies into two subtypes and these are now known by their names as the Fanaroff–Riley classes 1 and 2, or more briefly as FR 1 and FR 2 radio galaxies. The division is based upon the structures of the double radio sources using "the ratio of the distance between the regions of highest brightness on opposite sides of the central galaxy or quasar, to the total extent of the source measured from

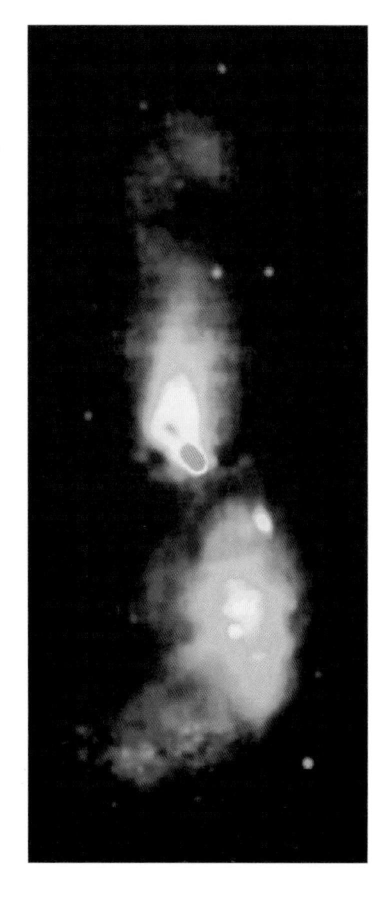

Figure 3.19 A 0.06 m (5 GHz) image of Cen A showing the outer parts of the radio lobes. The lobes are over 11° across as seen in the sky, over 20 times the size of the full moon. (Image courtesy of Norbert Junkes, Max Planck Institute for Radio Astronomy, Bonn, Germany.)

the lowest contour"[24]. The ratio is taken to be less than a half for FR 1 sources and greater than a half for FR 2 sources – in other words for FR 1s, the separation between the peak emission sources is less than half the total size of the radio lobes, while in FR 2s the separation is more than half the total size of the radio lobes. A consequence of the differing relative positions of the emission peaks within the radio lobes is that the lobes appear darker towards their edges for FR 1 objects while the radio lobes in FR 2 objects brighten towards their edges or ends. FR 2 objects are sometimes regarded as the classical double-lobed

[24]Fanaroff B.L., Riley J.M., Monthly Notices Roy. Astron. Soc., **167**, 31P, 1974.

radio galaxy. Cen A (Fig. 3.19) is an example of an FR 1 radio galaxy while Cyg A (Fig. 3.18) is an FR 2.

Perhaps surprisingly, the FR classification also correlates with the objects' radio brightnesses – at a wavelength of 0.17 m (178 MHz) FR 1 objects are fainter[25] than about $10^{25}\,\mathrm{W\,m^{-2}\,sr^{-1}}$ and FR 2s are brighter than this limit. FR 2s usually have a reasonably bright compact central radio source coinciding with the optical galaxy, but the central radio sources are much fainter in FR 1s, to the extent that they may be difficult to detect. The jets by contrast are usually more easily detectable in FR 1 galaxies than in FR 2s. Double jets are commonly found within FR 1 radio galaxies and single jets in FR 2s (Chap. 4) although at small scales (a few light years) single jets can occur in FR 1s as well. The jets in FR 2s are faster, sometimes being relativistic, than those in FR 1s and may be broken up into several "hot-spots". Their long-wave radiation is often strongly polarized. In the optical region, the central galaxies of both FR 1s and FR 2s are usually giant ellipticals with those for the FR 1s tending to be the larger – indeed an FR 1 elliptical is often the dominant galaxy in the cluster of galaxies to which it belongs. The optical spectra of the central galaxies of FR 2s often exhibit both broad and narrow emission lines, whilst those of FR 1s have just the narrow lines (cf. Seyfert 1s and 2s). They may thus also be classed as BLRGs and NLRGs (Sect. 3.2.2).

Some quasars, such as 3C47 (Fig. 3.16), have extended radio emission regions as well as their central AGNs. They may then receive an FR classification in addition to being classed as quasars – 3C47 is thus an FR 2 double-lobed radio galaxy as well as a quasar. However, quasars always have single jets and so are often classed as FR 2 objects even when they do not have significant extended radio emission. In a further link to other types of AGNs it is possible that if we were to see radio gal-

[25]The dividing brightness was calculated as $2 \times 10^{25}\,\mathrm{W\,m^{-2}\,sr^{-1}}$ in the original paper, but this was based upon a value for the Hubble constant of 50 km/s per Mpc. The figure quoted in the main text is based upon a value for the Hubble constant of 71 km/s per Mpc.

axies directly down the line of their jets, then FR 1s and FR 2s would appear to us as BL Lacs and HPQs respectively (Sect. 3.2.5).

It seems obvious that the jet that is moving in our direction should be feeding the radio lobe that is closest to us. However it is always nice to have observational support for such "obvious" conclusions. The Laing–Garrington effect provides that support. The effect is the observation that the radio emissions from the lobe associated with the brighter of two jets or with a single jet in a double-lobed radio galaxy is more strongly polarized than the radiation from the other lobe. The reason for the difference in polarization levels is that the more highly polarized lobe is closer to us so that its radiation passes through less of the material making up the galaxy's halo. The halo material depolarizes radiation passing through it[26] and so the more distant lobe has the lower levels of polarization.

In terms of the black hole model for AGNs (Sect. 3.2.1 and Chap. 5), double-lobed radio galaxies are clearly objects that are seen at high angles to the accretion disks' rotation axes. That is to say the rotation axes, and so also the jets, are close to the plane of the sky. The radio lobes then result when the material in the jets collides with the surrounding intergalactic medium. While many double-lobed radio galaxies have simple co-aligned linear structures similar to that of Cyg A (Fig. 3.18) others may have twisted jets, or more than two lobes, or other more complex structures. Multiple lobes may be attributed to there having been two or more outbursts of heightened activity around the central black hole. Twisted jets can arise in several ways such as the precession of the central black hole, or to there being two black holes in orbit around each other. The latter could easily arise after a merger between two galaxies during the interval before the black holes themselves merge into one. In a few cases, the jets are bent in the same direction so that they have a bow-shaped appearance. This seems likely just to be the result of the galaxy's individual motion through the local intergalactic medium.

[26]A process called Faraday rotation that is beyond the scope of this book.

4
Jets and Super-Luminal Motion

Summary

- The nature of jets in AGNs.
- Single jets in quasars and blazers.
- The causes of super-luminal motions.
- FR1 galaxies and double jets.
- The link between jets and radio emitting lobes.
- Boxes
  Relativistic beaming and boosting.

4.1 When Is a Jet Not a Jet?

The discussion of jets has been left to a separate chapter because of their importance in contributing to our understanding of what is happening inside AGNs. The reader will already have encountered frequent mention of jets and some of their properties in earlier chapters, here we bring the various phenomena, models and theories together to try and produce a coherent picture of what is happening.

In the context of AGNs, a jet is a collimated outflow of high-speed material. Outflows occur very commonly in a wide variety of astrophysical situations, but not all outflows are jets. The solar wind, for example, is highly supersonic but is not collimated since it is emitted radially from practically all over the solar surface. Similarly eruptions from Io's volcanoes are collimated, at least to begin with, but are moving only slowly. A more borderline case, although still not a genuine AGN-type jet, is that of the normal Sc spiral galaxy, M 101. This galaxy shoots out knots of material from its core at speed of around 100 km/s. The knots are emitted at about 11 million-year intervals from alternate sides of the galaxy. One suggestion to explain this phenomenon is that a large, but not super-massive, black hole orbits the galaxy's center and the knots are emitted each time the black hole passes through the gas-rich central parts of the core. The Whirlpool galaxy (M 51, see Fig. 7.1, below) on the other hand has jets that probably do fall on the AGN side of the definition. M 51 is classed as a weak Seyfert 2 or LINER galaxy. Early HST images of M 51's core show an edge-on torus of absorbing material about 100 ly (30 pc) across that is probably part of an accretion disk around a central massive black hole. Two high-speed jets emerge from the disk, but because the disk is perpendicular to the plane of the galaxy, they slam straight into the dense interstellar medium and thus produce only (relatively) small-scale ionized cavities close to the galaxy's center.

The requirements for jets to be collimated and moving at high speeds are not two separate conditions for an outflow to be defined as a jet because it is usually the high speed that leads to the collimation (unless the outflowing material is preventing from expanding sideways by some external means such as the pressure of a surrounding gas or by a magnetic field, see below). The material forming jets – usually a plasma of ions and free electrons – is hot, sometimes extremely hot, and so the thermal motions of its constituent particles are large. A mass of such material dumped into intergalactic space would simply expand in all directions. An unconfined sphere of plasma 10 AU (1,500,000,000 km) in diameter and at a temperature of 100,000 K, for example, would double in size in less than a year since its ions have an average speed of about 50 km/s[27]. Now, as already mentioned, unless there is some external force preventing or hindering the expansion, the material forming part of a jet is no different from the isolated sphere of plasma and it will expand in exactly the same way. For the jet to appear collimated to an outside observer, the velocity of the jet must therefore be many times the thermal speed of its constituent ions. The distance moved along the length of the jet in a certain time interval will then be much greater than the increase in its width arising from its expansion due to thermal motions.

If we define the opening angle of a jet as the angle made between the lines defining opposite outer edges of the jet, then a jet moving at 10 times the thermal speed of its particles would have an opening angle of about 12° and this would reduce to 1.2° and 0.12° (= 7.2 minutes of arc) at one hundred and one thousand times the thermal speed respectively. The opening angle never reduces to zero, but for most practical purposes it becomes undetectable, so that the jet appears to have parallel sides, at a jet velocity around a few hundred times the thermal speed, i.e. for jet speeds over about $0.1c$.

Some AGNs have relatively slow jets and these then tend to be diffuse and amorphous as would be expected when the thermal expansion of the plasma has become significant. Most AGNs however do have well collimated jets, implying high speeds for the material along the axes of the jets, with those speeds reaching relativistic values in many AGNs.

---

[27]The electrons travel much faster – over 2,000 km/s in this case – so it might seem that the expansion should be quicker. Indeed to begin with many electrons will zoom outwards ahead of the ions, but this will soon lead to a net positive electric charge on the main sphere of material and that will then hold back the electrons until they expand at the same rate as the ions.

Jets may be smooth and continuous, but are often broken up into a number of knots and concentrations, particularly at very small size scales close to the nucleus of the AGN. With Very Long Baseline Interferometry (VLBI) the innermost parts of a jet can be resolved down to mas[28] scales and the emergence of new knots from the core of the AGN followed. The birth of a knot is often accompanied by a flaring of the brightness of the core of the AGN and presumably is related to that outburst in some way.

Before moving on, we should clarify the differences between jets and beams. Jets, as just discussed, are streams of material. Beams are streams of radiation, like the emission from a lighthouse, although they can be made up of e-m radiation from any part of the spectrum. The radiation from jets is often beamed as a result of the relativistic motions of the material in the jets. The mechanism underlying relativistic beaming is discussed in Box 4.1.

The radiation from jets arises as synchrotron emission from relativistic electrons spiralling around magnetic fields. The loss of energy that this represents causes the electrons to slow down so reducing the intensity of subsequently emitted radiation. Eventually the electrons will slow to the point where synchrotron emission effectively ceases. One of the current problems with AGN models is that the lifetimes of relativistic electrons within jets seem to be too short (perhaps less than a century) to explain the jets' durabilities. The jets in NGC 6251 (Sect. 4.3) for example, are at least 650,000 ly (200 kpc) long, so that even if the jet material is moving at close to the speed of light, the electrons in the jets must retain sufficient energy to be able to continue to radiate synchrotron radiation for close to a million years. It seems likely that some process converts a fraction of the huge amount of the ordered kinetic energy of the jet represented by its movement through space into randomized motions of the electrons (and ions), thus re-accelerating the electrons whilst they are within the jet, but how this works is not yet

[28]Milliarcsecond – one mas is about the angular size of an astronaut on the Moon as seen from the Earth.

understood. The polarization of the synchrotron emission from jets indicates that the magnetic fields align with the long axes of the jets, and this may contribute to the jets' ability to retain their collimations over long distances and times.

4.2 Jets in Quasars and Blazars

The first jet belonging to a quasar was discovered simultaneously with the discovery of the first quasar. The lunar occultation observations of 3C273 (Sect. 3.2.4) that pinned down its position in the sky showed that it was a double radio source – one source turning out to be the center of the AGN, the other, its jet. More unusually, the optical identification of 3C273 also showed both the quasar and the jet (Fig. 3.14). 3C48, the first observed quasar, although not the first to be identified for what it was, was also later seen to possess a single jet.

Single jets have now been observed, usually at radio wavelengths, for many quasars. However those quasars that also have radio lobes (FR 2 radio galaxies, see Sect. 3.2.6) invariably have two lobes on opposite sides of the AGN. One lobe is almost always aligned with the visible jet and is thought to be formed through the interaction of that jet with intergalactic material (Chap. 3). It is to be expected however from the unified model of AGNs (Chap. 5) that two jets will be produced moving in opposite directions along the rotational axis of the central black hole. The existence of the second radio lobe is evidence that just such a second jet does exist. In many other AGNs, two jets are indeed detectable. However the counter-jet in quasars has almost never been found. Only for one object – a radio source in Libra (B1524–136) that has been identified with a 20^{m} optical source and tentatively classed as a quasar – has a counter-jet clearly been picked up. Even here though and even if the object is a true quasar, the physical situation must differ markedly from that of a "normal" quasar since the counter-jet is not directly opposed to the main jet (i.e. at 180° to it) but is shooting off to the side at an angle of about 130°.

Suggestions to explain the lack of a counter-jet in quasars have included the counter-jet being intrinsically fainter than the main jet (for some unknown reason), absorption of the emissions from the counter-jet by gas and dust clouds in the central part of the AGN and the jet truly being single, but switching its direction through 180° at regular intervals, so that two radio lobes are produced. The latter explanation has been dubbed the "flip–flop" model and suffers from the lack of any known means of producing single jets of the required energy and of sufficient mass except under temporary and unusual circumstances (for example a double jet that happens to have one jet impeded by a large gas cloud). There is also no credible explanation for the jet's reversal of direction. Thus it is now almost universally accepted that there are two jets in quasars and that the reason for the invisibility of the counter-jet lies in the phenomena of relativistic beaming and boosting (Box 4.1). This results in the emission from the jet that is approaching us being concentrated along the direction of motion of the material in the jet and its intensity increased markedly within the beam over what would be its isotropic emission. The counter-jet likewise has its emissions concentrated into a beam along its direction of motion, but since that is directed away from us, only the very much weakened rearward emissions are coming in our direction. Thus if the jets are angled at 10° to the line of sight and their material is moving at 95% of the speed of light, then the approaching jet will be several hundred thousand times brighter than the receding jet. In most cases such a faint counter-jet would then be buried in the background noise of the radio telescope and so not detected.

Box 4.1 Relativistic Beaming and Boosting

Imagine that you are driving at night through a snowstorm. If there isn't much wind then you know that the snowflakes are falling vertically, however when you see them shining in the beams of your headlights, then they seem to be coming at you from somewhere above and ahead of you (Fig. 4.1). This is because you are moving

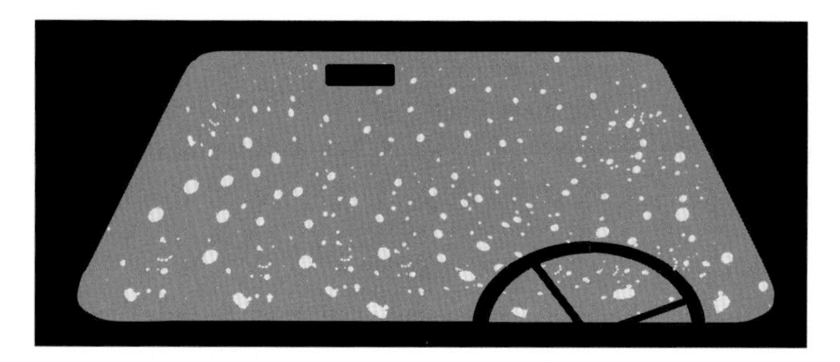

Figure 4.1 Driving through a snowstorm at night – the vertically falling snowflakes appear to be coming towards you.

horizontally along the road while the snowflakes are coming straight down, so the *relative* velocity between you and the snowflakes is such that they are coming towards you from a point that lies somewhere on the line connecting your direction of movement with that of the snowflakes. If you drive faster (this is just a *thought* experiment – don't try it in practice!) then the point from where the snowflakes seem to come will move closer to your direction, i.e. they will approach from somewhere nearer to directly ahead of you. If you stop, then the snowflakes will fall from the zenith.

The phenomenon just described for a snowstorm occurs whenever objects are in relative motion and is called aberration. The same effect forces someone hurrying through a rain shower to angle their umbrella forward, a hunter to "lead the bird" when aiming a shotgun and results in the meteors within a meteor shower appearing to originate from the radiant.

In the cases of aberration just considered, the relative velocity between the two objects is the vector sum of their individual velocities. Thus if you were driving at 30 kilometers per hour (19 mph) and the snowflakes were falling at 10 km/h (6 mph), then the relative velocity would be 31.6 km/h (20 mph) from a point 18° above the direction in which you were moving. The relative velocity is thus larger than either of the two individual velocities. When we come to consider what happens when it is photons (light) that are falling onto a moving object, the situation becomes slightly different because we know from special relativity that the speed of light is constant. If we imagine the snowflakes being replaced by photons, then the photons are moving downwards at the speed of light, and also have a relative velocity with respect to you that is equal to the speed of light. This apparently contradictory result has been verified experimentally many, many times, and is the foundation upon which special relativity is based, the reader curious to know more about the effect is referred to more specialist texts (Appendix 1). Despite there being no change between the magnitudes of the initial and relative velocities of the photons in this situation though, there is still a change in the direction from whence they seem to be coming.

Taking a more relevant astronomical situation, we observe objects in the sky from an Earth that is moving at 30 km/s around its orbit. Let us imagine that the Earth's orbital velocity is reduced to zero momentarily and during that instant an astronomer centers a star in the eyepiece of his or her telescope. What will he or she see when the Earth resumes its normal orbital motion? Unless the star is directly ahead of or behind the Earth, then it will move off-center in the eyepiece by a small angle. In order to center the star again, the astronomer will need to move the telescope through that small angle in a direction towards where the Earth is traveling at that moment. Like the apparent direction of the snowflakes from a moving car, the apparent direction of the photons from the star (i.e. the direction in which we have to point the telescope in order to see the star) moves towards the direction in space for which the Earth is heading.

We may see how this aberration of light arises by considering what we mean by "centering the star in the eyepiece". Assuming that we have a good telescope – no misaligned or poor quality optics, etc. – then when the star is seen to be in the center of the eyepiece, a photon that passes through the center of the telescope's objective lens or is reflected from the center of its objective mirror will also pass through the centers of the lenses making up the eyepiece. Thus if the star is centered in the telescope when the Earth is stationary, then when the Earth is moving and carrying the telescope with it, a photon passing through the center of the objective will no longer pass through the centers of the eyepiece's lenses. This is because during the (very short) interval of time that it takes the photons to travel down the telescope from the objective to the eyepiece, the Earth's motion will have moved the eyepiece and it will no longer be in a position for the photons to pass centrally through it (Fig. 4.2). Restoring the star to the center of the field of view means tilting the telescope towards the Earth's direction of motion at that instant. Since the Earth's orbital speed of 30 km/s is 1/10,000 of the speed of light, the maximum angle by which the telescope will need to be adjusted to compensate for aberration is 1/10,000 of a radian, i.e. by 20.5 seconds of arc at most.

Now the direction of the Earth's orbital motion rotates through 360° over a year, and so the direction in which the telescope needs to be moved to correct for aberration also rotates over the same

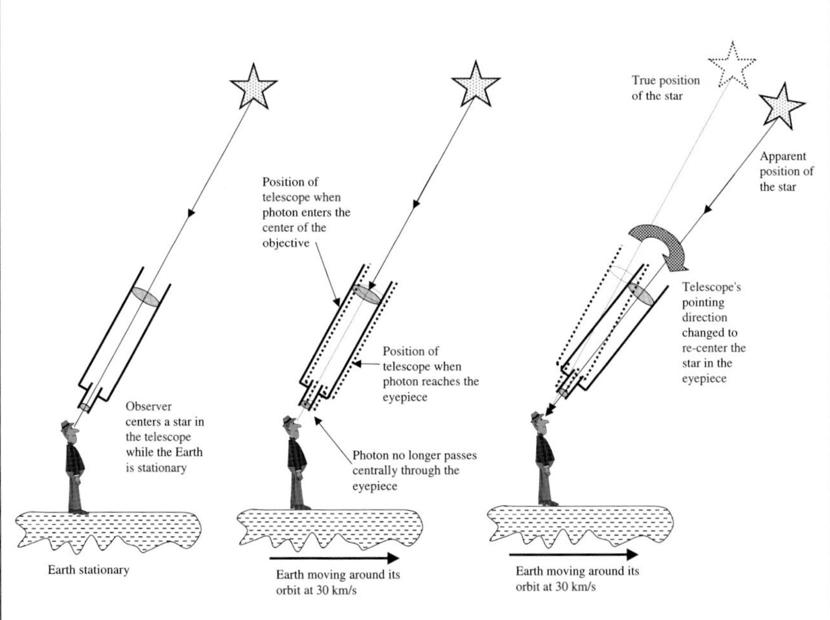

Figure 4.2 The aberration of starlight.

interval. Changing the observational situation slightly to where we do not move the telescope to correct for aberration, then over a year we would see the star's position trace out an ellipse whose major axis was 41 seconds of arc long. The shape of the ellipse depends upon the position of the star with respect to the plane of the Earth's orbit (i.e. the ecliptic). Directly above or below the ecliptic (in Draco and Dorado), the stars move in small circles with 41 second of arc diameters. The nearer a star is to the ecliptic, the narrower its aberrational ellipse becomes, until stars on the ecliptic just move back and forth in a straight line. An angle of 20 seconds of arc is small, but easily measurable, so that the aberrational movement of stars was discovered by James Bradley as early as 1728 and was the first concrete proof that the Earth was really moving through space as Copernicus' model of the solar system required.

The Earth's orbital motion is small compared with the speed of light, and so the angle of aberration is also small. However imagine that the Earth were moving a thousand times faster, at 30,000 km/s around its orbit. The aberrational shift in stars' positions would then be by nearly 6°. A careful examination of the sky would also show that it was unbalanced – more objects would be found within the observed hemisphere centered on the Earth's direction of motion than in the trailing hemisphere. The imbalance arises because all objects in that part of the sphere centered on the Earth when it is stationary (or moving slowly) that are within angles to the Earth's direction of motion from 0° to 96° would be moved into the forward hemisphere when the Earth is moving at 30,000 km/s. Only those within angles from 96° to 180° would now appear visible within the trailing hemisphere. If objects in the sky were initially distributed isotropically, then from the 30,000 km/s Earth, 55% would now be seen in the forward half of the sky and 45% in the trailing half.

Of course, in imagination it is just as easy to picture the Earth orbiting at 60,000 km/s or 100,000 km/s, etc., but at velocities as high as these the simple picture that we have developed so far becomes distorted by the effects of relativity, leading to the aberration angles being larger than might otherwise be expected. Thus taking a star that was originally to be seen (when the Earth was moving slowly) at 90° to the direction of motion, then at 100,000 km/s, the simple picture suggest that it should shift towards the direction of motion by an angle of about 18.5°, however taking account of special relativity, the shift in apparent position becomes about 19.5°. The aberration angle for our object that was originally at 90° to the velocity direction continues to increase with velocity, reaching 90° at the speed of light (Table 4.1). The imbalance between the numbers of

Table 4.1 Effect of aberration on an isotropic initial distribution of stars.

Velocity (km/s)	v/c	Aberration angle for an object seen at 90° to the velocity direction when the observer is stationary. (°)	Percentage of the whole sky that is visible when stationary which is now concentrated into the forward hemisphere[29]
0	0	0	50
50,000	0.167	9.6	58.3
100,000	0.333	19.5	66.7
150,000	0.5	30	75
200,000	0.667	41.8	83.3
225,000	0.75	48.6	87.5
250,000	0.833	56.4	91.7
275,000	0.917	66.4	95.8
280,000	0.933	69	96.7
285,000	0.95	71.8	97.5
290,000	0.967	75.2	98.3
295,000	0.983	79.5	99.2
296,000	0.987	80.6	99.3
297,000	0.99	81.9	99.5
298,000	0.993	83.4	99.7
299,000	0.997	85.3	99.8
300,000	1	90	100

[29]This is equal to the percentage of the total (isotropic) emission from a moving object that is seen by an outside observer being emitted into the forward direction.

objects to be seen in the forward and trailing hemispheres also continues to increase until at 90% of the speed of light 95% of the objects visible in the sky would now be the forward half of the sky and only 5% in the trailing half and at 99% of the speed of light those figures become 99.5% and 0.5% (Table 4.1).

So far we have considered the effect of aberration upon what might be seen from the Earth when traveling at improbably high speeds. The calculations however are unchanged whether the photons are traveling in, towards the Earth or outwards, away from the Earth. Thus replacing the Earth in our imaginations with a more astrophysically realistic high-speed electron, then if that electron is emitting synchrotron radiation, the outward traveling photons from it will be affected by aberration in the same way as those coming in towards an observer on the Earth would be. If the electron's original emission is isotropic[30], an outside observer will see the emission concentrated into the forward direction (i.e. the same direction as that in which the electron is traveling). Most of the radiation will be emitted into a beam whose angular radius is roughly 90° minus the aberration angle corresponding to the electron's speed. Thus for an electron moving at 95% of the speed of light, 97.5% of its radiation will be emitted into the forward direction with most of this focused into a beam about 36° across (Table 4.1). At 99% of the speed of light, the beam contracts down to a width of 16° and 99.5% of the electron's radiation is emitted into the forward direction.

Finally, then, we have arrived at the idea of relativistic beaming. When material (almost always a jet of free electrons in astrophysical situations) is moving at relativistic speeds, its radiation is concentrated into a narrow beam that is emitted along the direction of the material's motion. An observer aligned with the beam will see intense emission (which if it is assumed to be isotropic will result in a greatly

[30]The electron actually has a dipolar (doughnut-shaped) emission pattern, but this will not affect the discussion greatly.

overestimated value for the object's true luminosity). Observers to the side will see much lower intensities while those behind the beam will be able to see little or nothing.

The effect of aberration is intensified by relativistic boosting of the beam from the approaching jet and the fading of the radiation from the receding jet. Relativistic time dilation means that time passes more slowly for the material forming a high-speed jet than it does for ourselves. We therefore receive the energy emitted over a certain interval by the jet over a longer time period and so the effective observed intensity of the emitted radiation is reduced. However in our own frame of reference the material forming the jet is traveling close to the speed of light. Thus following the same line of argument that is used to explain super-luminal velocities within the main text, light emitted towards us is being followed closely by the material that emitted it. Thus considering the emission from material moving towards us at 90% of the speed of light over a period of one second, the material will be 270,000 km closer to us when it emits radiation at the end of that second than it was when it emitted radiation at the start of the second. The radiation emitted by the material over one second in our time frame is thus received by us in a tenth of a second – and so the effective observed intensity of the emitted radiation is increased. The net effect of these two processes is to increase the observed intensity of the radiation emitted by the material forming the approaching jet. The actual value of the increase depends upon the nature of the spectrum of the jet's emission, but for a jet approaching us at an angle of 10° to the line of sight and at a velocity of $0.9c$, the intensification is by a factor of around 50–60. Radiation from the receding jet is reduced in intensity somewhat similarly, so the approaching jet appears to be about 50^2 or 2,500 times brighter than the receding jet. Boost factors for other velocities and angles to the line of sight are shown in Table 4.2.

These two factors together mean that normally the approaching jet is far brighter than the receding jet. Thus in most cases (possibly

Table 4.2 Relativistic boosting of jets' intensities.

	Angle of jet's track to the line of sight = 0°		Angle of jet's track to the line of sight = 10°		Angle of jet's track to the line of sight = 20°	
v/c	Boost factor (approx)	Brightness of approaching jet compared with that of the receding jet	Boost factor (approx)	Brightness of approaching jet compared with that of the receding jet	Boost factor (approx)	Brightness of approaching jet compared with that of the receding jet
0	1	1	1	1	1	1
0.25	2.2	5	2.1	4	2	4
0.5	5.2	30	5	25	4.4	19
0.75	19	350	16	250	11	120
0.8	27	700	23	530	14	200
0.85	43	1,800	34	1,200	18	320
0.9	80	6,000	56	3,100	23	530
0.95	240	60,000	110	12,000	25	630
0.99	2,800	8,000,000	180	32,000	8.3	69
0.999	90,000	8,000,000,000	21	440	0.4[31]	0.16

[31]Relativistic time dilation is starting to predominate over relativistic pulse compression here. The approaching jet however is not actually likely to be fainter than the receding one because of beaming and because although electron speeds can reach 99.9% of the speed of light, the bulk motion of the material forming the jet is likely to be considerably less than this.

Table 4.2 *Continued*

	Angle of jet's track to the line of sight = 30°		Angle of jet's track to the line of sight = 40°		Angle of jet's track to the line of sight = 50°	
v/c	Boost factor (approx)	Brightness of approaching jet compared with that of the receding jet	Boost factor (approx)	Brightness of approaching jet compared with that of the receding jet	Boost factor (approx)	Brightness of approaching jet compared with that of the receding jet
0	1	1	1	1	1	1
0.25	1.9	3.6	1.7	2.9	1.5	2.3
0.5	3.6	13	2.8	7.8	2.1	4.4
0.75	6.7	45	3.8	14	2.1	4.4
0.8	7.5	56	3.7	14	1.9	3.6
0.85	8	64	3.4	12	1.6	2.6
0.9	7.7	59	2.8	7.8	1.1	1.2
0.95	5.5	30	1.5	2.3	0.5	0.3
0.99	1	1	0.2	0.04	0.06	0.004
0.999	0.04	0.0002	0.007	0.00005	0.002	0.000004

all) where a single jet is observed within an AGN, it is not because there is only one jet, but because the jet approaching the observer has its luminosity enhanced enormously by the relativistic beaming and boosting effects, while the receding jet similarly has its luminosity reduced until it is below the detection threshold.

Independent confirmation that some AGN emissions must be beamed and boosted lies in the observation that the apparent luminosities of a number of AGN cores imply temperatures exceeding 10^{12} K. Temperatures that high should lead to intense x-ray emissions, but these are not observed. The temperature of the core must therefore be far lower than 10^{12} K and so its luminosity must also be less than it appears to be. The lower luminosity is easily achieved if the core is not emitting isotropically while beaming intensifies the emissions coming our way.

Early radio telescopes had very poor resolutions by the standards of optical astronomy and special techniques – such as using lunar occultations – were needed to pin down the first quasar positions. Radio astronomers however quickly started using two or more individual radio aerials to form interferometers with much improved resolutions. In an interferometer, the resolution depends upon the distance between the aerials, not the sizes of the aerials themselves. In the quest for better resolution, the sizes of radio interferometers quickly rose from the half-mile instrument at the Mullard radio observatory in Cambridge to the Very Large Array (VLA) in New Mexico with its 27 dish-type aerials and maximum separation of 36 km. Even greater separations were obtained with instruments like MERLIN (Multi-Element Radio-Linked Interferometer Network) based at Jodrell bank but with its most remote telescope at Cambridge giving it a 217 km baseline. Operating at 5 GHz (60 mm wavelength) MERLIN has a resolution of 50 mas – slightly better than the optical resolution of the Hubble space telescope. The ultimate in interferometer sizes, at least on the Earth, has been reached with VLBI. Here the individual aerials are separated by thousands of kilometers. Their outputs can no longer be combined directly as in smaller interferometers but are recorded along with timing pulses and the recordings combined at some later time.

With VLBI, resolutions of a single milliarcsecond or better can be reached, but the field of view is very small. VLBI is thus of little use for observing angularly large radio sources such as the lobes of radio galaxies, but it is ideal for studying the compact central regions of AGNs. When VLBI was first used to look into the central regions of quasars in the late 1960s, it was found that there were often several radio sources present whose relative positions sometimes changed with time. Knowing the quasar's distance it is an easy matter to convert the angular motion observed by VLBI into the physical motion through space of the radio sources. In 1971, observations of the OVV quasar 3C279 in Virgo revealed a startling phenomenon – the radio sources were moving apart at three and a half times the speed of light (Fig. 4.3). Soon similar "super-luminal" motions were found in other quasars, including 3C273, and also in every blazar that was studied. In some cases the speeds reached nearly thirty times that of light.

Now a century of special relativity has deeply ingrained the idea in the minds of physical scientists that the speed of light in a vacuum is an upper limit to the speed of any material object through space. The observation of objects apparently moving much faster than the speed of light thus came as a severe shock. Fortunately for the sanity of a generation of physicists and astronomers, Sir Martin Rees had predicted the possibility of apparently faster than light motion as early as 1966 – four years earlier than its discovery in practice. Before however examining in more detail how super-luminal motions actually arise in AGNs, let us take at a related but simpler situation that also gives rise to such velocities and which may help to understand what happens in AGNs.

Imagine somewhere and somewhen within the universe that an attempt is to be made on the galactic land speed record, then standing at 265,000 km/s. On a suitably large planet a distance of 270,000 km is measured along a smooth beach. Flash lamps are placed at each end of

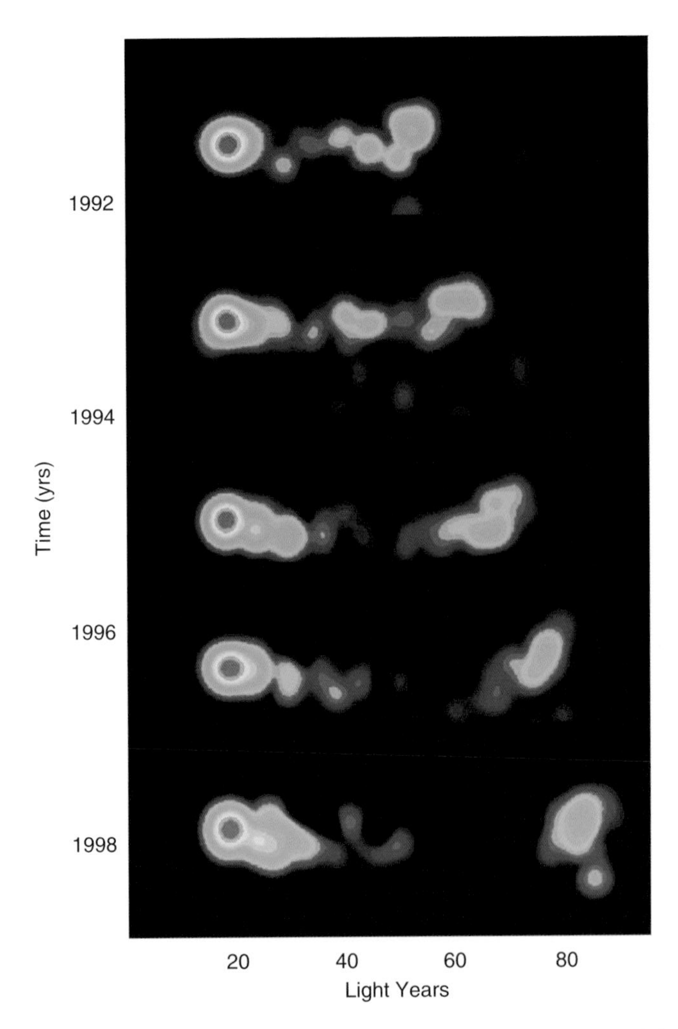

Figure 4.3 Super-luminal motion observed by VLBI in 3C279 at 3.5**c**. (Image courtesy of NRAO/AUI, Anne Wehrle and Glenn Piner.)

the measured distance and so arranged that they will go off when the car passes by them. Officials are placed to the side of the track and at its ends to measure the intervals between the flashes. The record attempt is made and as usual two runs are completed in opposite directions along the measured distance. When the timings by the officials are inspected, official A, standing to the side of the track measured exactly one second between the flashes on both occasions – a speed of 270,000 km/s and a new galactic record. Unfortunately the officials standing in line with the track completely disagreed with this measurement. On the first run official B, towards whom the car was heading, came up with an incredible 2,700,000 km/s – a super-luminal velocity nine times that of light. Official C, at the other end of the track, on the other hand found that the car only reached a paltry 142,000 km/s. However on the second run by the car over the measured distance, the two in-line official timers found that their measurements were reversed – C now measuring the speed to be 2,700,000 km/s and B getting 142,000 km/s.

So how can a single car making two runs over the same measured distance do so at three different speeds, one of which was highly super-luminal? The answer lies in the car's speed being close to, but less than, the speed of light. When the car passed the first flash bulb, the flash that it emitted expanded outwards in all directions at the speed of light, 300,000 km/s. The car, however, was moving at 270,000 km/s; hence those photons emitted along the measured distance, and so towards the in-line timing official B, were gaining on it by only 30,000 km/s. When the car reached the second flash bulb, one second after passing the first, the photons from the first flash were only 30,000 km ahead of those from the second flash. The official B therefore measured the two flashes as separated by just a tenth of a second – and a distance of 270,000 km covered in 0.1 seconds gives an apparent speed nine times that of light! The car however was moving away from official C at 270,000 km/s during its first run. From official C's point of view, therefore, when the second flash was emitted it was 570,000 km (= 300,000 km, i.e. the distance traveled by light in one second, +270,000 km) behind the first flash. Hence official C noted 1.9 seconds between the two flashes and so recorded an apparent speed of 142,000 km/s. The situations of officials B and C were reversed during the car's second run and so their results were also switched over. Thus we see that apparently super-luminal speeds can occur when an object traveling at a high, but *subluminal* speed, does so along a track that is directed towards the observer.

The salutary example of the land speed record attempt shows that bizarre things can happen when you are observing something that is approaching you at less than, but close to, the speed of light. The superluminal velocities observed in AGNs also arise because material moving at relativistic velocities is traveling towards us. To begin with though, the observations seem to be exactly the opposite of what happened during the land speed record attempt – the material in AGNs appears to be moving faster than the speed of light *across the line of sight*. It would seem therefore, that we are in the position of the timing official A at the side of the measured distance and so should be determining speeds correctly. However as Sir Martin Rees realized, we are actually not equivalent to any of the observers in the land speed record attempt, but are somewhere in between their positions – probably nearer to the position of the in-line timing officials B and C, but still to one side of the direction of motion of the material. Also we do not have a convenient measured distance with flash bulbs to make the reckoning easy, but must carefully ascertain the angular separation of a moving "bullet" of material from the center of the AGN and combine it with the distance of the AGN in order to obtain the physical separation of the "bullet" from the AGN.

Suppose then, that a bullet of material has been fired off from the center of the AGN at 90% of the speed of light and that we are 10° to one side of its track. Because the material is not approaching us directly, there is a component of its velocity amounting to 0.156*c* (46,800 km/s) across the line of sight. Consider now the two actual positions for the "bullet", separated by one second of time. During that time interval the "bullet" will move 46,800 km across the line of sight. Along the line of sight though, like the situation in the land speed record attempt, the bullet of material will be moving close behind the light that it has emitted. The light emitted at the end of the second will thus be trailing only 34,100 km[32] behind the light emitted at the start of the second. If we are monitoring the AGN continuously, then we will obtain images from the light emitted by the bullet at the start and end of the second that, to us, are separated by just 0.114 s (cf. Fig. 4.3). From our point

[32]Not 30,000 km as in the case of the land speed record attempt because the material is no longer coming straight towards us.

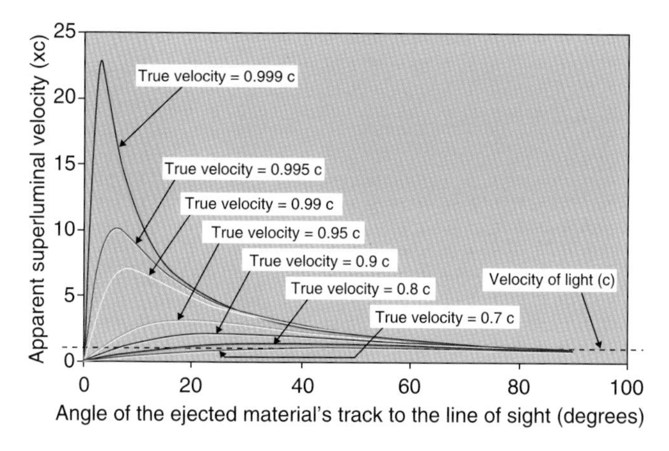

Figure 4.4 The variation of apparent super-luminal velocity with the true velocity of the ejected material from an AGN and the angle of its track to the line of sight[33].

of view, therefore, the "bullet" has moved 46,800 km across the line of sight in 0.114 seconds of time giving 411,000 km/s, a speed 1.4 times that of light.

An apparent super-luminal velocity of $1.4c$ across the line of sight thus results from material that is actually traveling at the *subluminal* velocity of $0.9c$ and which is on a track 10° to the side of our line of sight. The genuine observational situation uses VLBIs and so operates at radio wavelengths rather than light, and the time separation between the radio "images" is likely to be months, if not years, rather than 0.114 seconds used above, but this does not change the analysis in any significant way. Higher and lower super-luminal velocities result from other inclinations of the ejected material's track to the line of sight and for other values of its subluminal velocity and these are illustrated in Fig. 4.4. From the unified model for AGNs (Chap. 5) we might expect about one in twelve of those AGNs that have sufficiently high-speed jets to

[33]This graph shows zero velocity whenever the material is coming directly towards us – in apparent contradiction of the value ($9c$) obtained in the galactic land speed record scenario. But there is no contradiction – the graph gives apparent velocities *across* the line of sight, the calculation in the land speed record case was for apparent velocities *along* the line of sight.

show super-luminal motions in excess of about five times the velocity of light. The number of AGNs with measured super-luminal motions however is currently too small to provide any statistically significant test of this prediction.

Thus the super-luminal motions observed in quasars and blazars are direct evidence that these objects contain jets of material moving at relativistic speeds in our direction. The fact that all blazars so far studied have super-luminal motions whilst only some quasars do so, is further evidence that our line of sight is almost directly along the track of the jet in blazars, but is inclined at a smallish but larger angle to the track in quasars. Relativistic beaming and boosting (Box 4.1) mean that the emissions from an approaching jet are likely to dominate those from other parts of the AGN. This is especially the case for blazars, where the jets' emissions are almost the only detectable radiation. In these circumstances, the intrinsic luminosity of the object will be greatly overestimated unless the beamed and boosted natures of the emissions are taken into account.

4.3 Jets in Other Types of AGN

Since super-luminal motions depend upon both high velocity and upon that velocity being directed towards us, it is not surprising that they are not found in objects, such as Seyfert 2 and FR 1 galaxies, where the jets are thought to be moving more or less across the line of sight. However there is one notable exception to this rule – the giant elliptical galaxy, M 87, at the center of the Virgo cluster of galaxies is classed as an FR 1 double-lobed radio galaxy and also known as the radio source Vir A. It has a single jet (Fig. 3.2) that is bright enough at optical wavelengths to have been discovered in 1918 by Heber Curtis at the Lick observatory and which can easily be imaged today by an amateur astronomer using a small telescope and a CCD camera (Chap. 7). The jet also emits strongly in the infrared, ultraviolet and x-ray regions, as well, of course, as at radio wavelengths. M 87 is amongst the closest radio galaxies to us at a distance of around 60 Mly (19 Mpc) and so has been studied in great detail. Its jet is some 6,000 ly (2 kpc) long and it can be imaged at radio wavelengths in its inner region down to scales of a couple of light-weeks (0.01 pc). The origin of the jet appears to be no more than 0.03 ly (0.01 pc) from the core of the AGN. In that inner region super-luminal velocities up to 6c are found for parts of the jet. Even for M 87 though, where the inclination of the track of the jet to the line of sight may be larger than is the case for quasars and blazars, a counter-jet has yet to be detected with certainty, although a VLBI image obtained in 2003 at a frequency of 86 GHz (3.4 mm wavelength) and a resolution of 0.06 mas does show a faint feature that could be the counter-jet or could equally well be the core of the AGN.

More generally FR 1 radio galaxies tend to have both jets detectable, while FR 2s just show a single jet. The FR 1 radio galaxy in Ursa Minor, NGC 6251 for example, has a spectacular main jet that has led to it being named the "blowtorch galaxy" (Fig. 4.5). The main jet is highly collimated and observable from scales of a degree (6 Mly, 2 Mpc) down to milliarcseconds (2 ly, 0.6 pc). There is however a much fainter counter-jet to be seen at wavelengths of a few tenths of a meter (1 to 2 GHz). Cyg A (3C405) however, which is an FR 2 radio galaxy, has also had a faint counter-jet detected at milliarcsecond scales, so the division between the two types is not completely clear-cut.

The jets in FR 1 galaxies are usually brighter, relative to the radio lobes, than in FR 2s and tend to be unbroken filaments. The jets in FR 2 galaxies by contrast are often broken up into a number of "hot spots", their velocities tend to be higher and their radio emissions are more strongly polarized. The material forming the jets in both types of FR galaxies is of lower density than the surrounding gas and this may help the jets retain their collimation. Velocities in the jets are supersonic with Mach numbers of 10 or more for the FR 2 galaxies and a quarter of that or less for the FR 1 galaxies. The jets in FR 1 galaxies often seem to undergo a transition to sub-sonic speeds after traveling a few thousand light years (kpc), possibly at points where they encounter denser parts of the interstellar or intergalactic medium. After such a transition

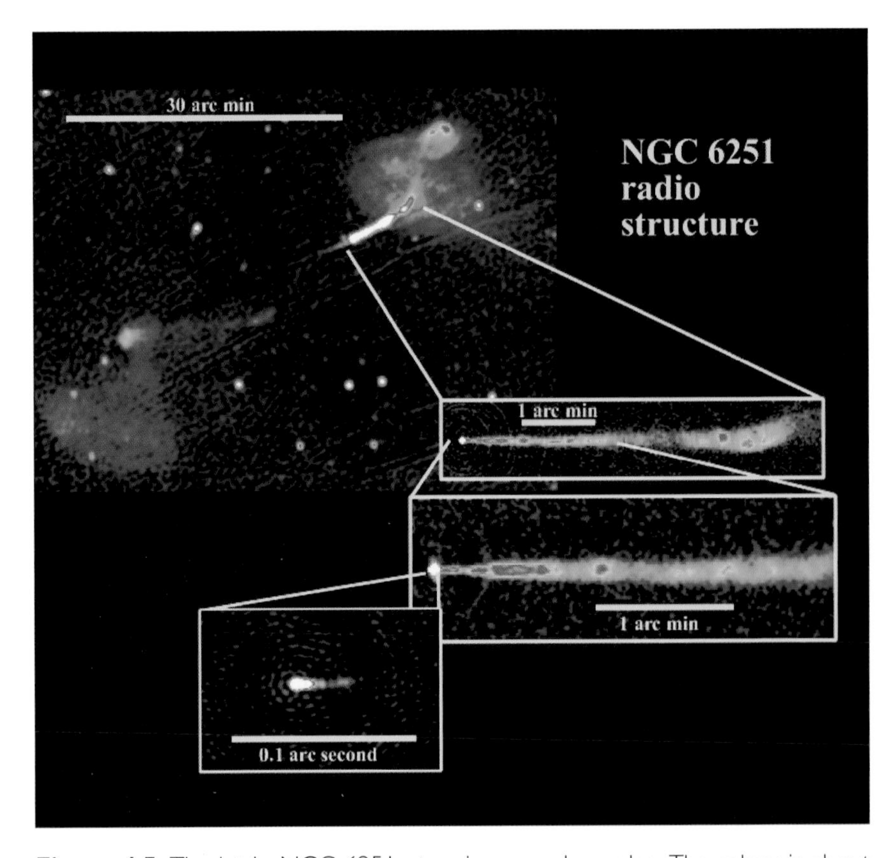

Figure 4.5 The jet in NGC 6251 at various angular scales. The galaxy is about 330 Mly (100 Mpc) distant so that one second of arc corresponds to about 1,600 ly (500 pc) and one minute of arc to about 100,000 ly (30 kpc). The counter-jet is just visible on the largest scale image, color-coded blue/green and to the left of the main jet. The streaks and circles visible on some of the images are artefacts produced by the interferometers used for the observations. (Images courtesy of William C. Keel, Karl-Heinz Mack, Rick Perley and Dayton Jones. Some of the data was originally published in *Astrophysical Journal*, **305**, 684, 1986 (reproduced by permission of the AAS) and *Astronomy and Astrophysics Supplement*, **123**, 423, 1997.)

to slow speeds, the jets lose collimation and become turbulent and chaotic. The differences between FR1, FR 2 galaxies and their jets seem to be closely linked. For a given jet radio luminosity, an FR 1 galaxy will generally be significantly brighter and more massive than an FR 2 galaxy. It is thus possible that the giant elliptical galaxies hosting FR 1 radio sources contain more gas and dust in their inner regions and this acts to slow down their jets compared with the FR 2s where the jets are freer to blast unhindered out of the host galaxy.

In a number of AGNs the jets appear to have changed direction on one or more occasions. There are several possible explanations for such a phenomenon that are based upon the central black hole of the AGN (Sect. 3.2.1 and Chap. 5) actually being a pair of black holes and forming a binary system. Such a scenario is not as unlikely as it might seem and could result from the collision and merger of two galaxies, each of which had a single central black hole before the interaction. The jets from a binary black hole might then change direction due to the orbital motions of the black holes, or might be emerging along the rotational axis of one of the black holes, whose direction in space is then precessing through the gravitational effects of the second black hole. A helically-shaped jet such as might be produced by the effects of precession could appear to change direction in a discontinuous manner if only some parts of its track are correctly aligned towards the observer for its luminosity to be enhanced by relativistic boosting. The radio galaxy 3C75 has two FR1-type nuclei separated by about 25,000 ly (7.5 kpc) both with twin jets (Fig 4.6) which will presumably will merge sometime in the future.

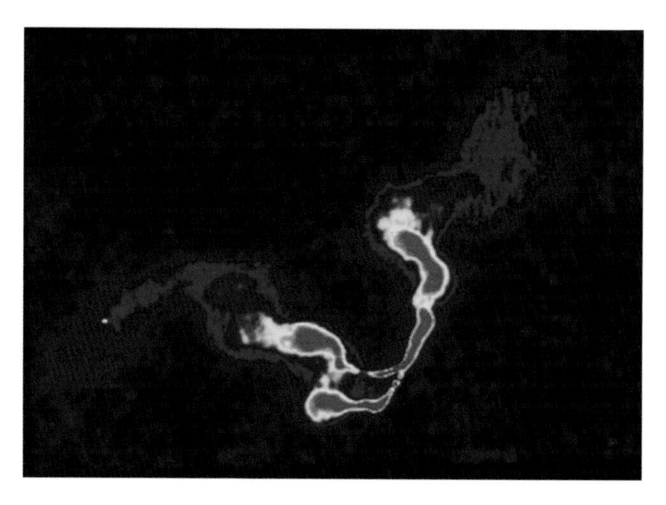

Figure 4.6 A 0.2-m (1.4 GHz) radio image of 3C75 showing its double FR I nuclei and four jets. (Image courtesy of NRAO/AUI, F.N. Owen, C.P. O'Dea, M. Inoue, J. Eilek and P. Smiley.)

Some 7% of radio galaxies have jets that are X-shaped, i.e. their direction has changed by close to 90°. A possible explanation for such a drastic change could lie with the collision and merger of two black holes. Computer modeling suggests that a large enough flip of the spin axis would occur even if one black hole was five times more massive than the other. The quasar 4C.73.18 is perhaps the most likely known candidate for containing a binary black hole. Waves in its jet suggest a period for the black hole binary of three years with individual masses for the black holes of about 100 megasuns. If this model is correct, the black holes are likely to collide and merge in about one million years from now.

Finally double jets that both bend in the same direction, leading to a bow-shaped structure, seem likely just to be the result of the AGN moving through the local intergalactic medium whose dragging effect then causes the jets to lag behind the galaxy.

When they are detectable at all, the jets in radio-quiet AGNs generally seem to have lower speeds and to be more diffuse than those in radio-loud objects – indeed differing jet speeds are widely held to be the reason why some AGNs are radio-loud and others radio-quiet. Some Seyfert galaxies, although classed as radio-quiet (see M 51 discussion earlier), are found to have small and weak radio emitting jets.

4.4 Jets and Radio Lobes

The radio-emitting lobes of AGNs are often linked to the central galaxy by jets. When no jet can be found leading to a radio lobe, then almost certainly there is a jet present, but it is too faint to detect. The jet's motion in such cases is away from us and relativistic beaming and boosting (Box 4.1) reduce its brightness in the rearward direction until it is buried in the background noise. It seems obvious that jets are transferring energy and material from the central black hole of the AGN to the radio lobes thus both producing the lobes and providing their continuing sources of energy.

The jets, especially the faster ones, contain a considerable amount of kinetic energy – far more than is emitted as radiation by the jet. The interstellar and halo material of the galaxy hinders the passage of the jet only slightly so that it still has most of the kinetic energy with which it started when it arrives at the radio lobe. There it encounters denser material and is suddenly decelerated, forming a shock front. Within the shock front the ordered kinetic energy of the jet is converted into randomized relativistic motions of the electrons and ions forming the material of the jet. The jet will also have carried a magnetic field with it – indeed some models for jets explain their pencil-like collimation through the effects of magnetic fields (Sect. 4.1). The combination of relativistic electrons and magnetic fields leads to the emission of synchrotron radiation and the production of the brightness peak in the radio lobe. The material of the jet gradually diffuses out from the brightness peak and forms the remainder of the radio lobe. In the larger and brighter lobes, the total amount of energy in the form of magnetic fields and high-speed particles can amount to the equivalent of the complete conversion of 10 megasuns into energy.

5 The Unified Model for AGNs

Summary

- Black holes – some of their properties and the phenomena associated with them.
- Evidence that black holes are the central objects within galaxies and AGNs, and not some other form of concentrated mass.
- Energy generation via black holes – the Penrose process and accretion disks.
- Minimum (Eddington) masses for sources of a given luminosity.
- The unified model for AGNs – its successes and current problems.

5.1 Black Holes

5.1.1 What Is a Black Hole?

It would be easy to write a book the size of this one, or perhaps several such books, just on black holes alone. Here therefore we shall review only their basic properties and look at those aspects of black holes that are particularly relevant to AGNs.

The notion of an object whose mass is such that the escape velocity from its surface equals or exceeds the speed of light dates back over two centuries. The Reverend John Michell, an English geologist, suggested the idea in a paper to the Royal Society in 1783. However it required Einstein's 1915 theory of general relativity to put the concept onto a sound theoretical basis. According to general relativity, the fabric of the universe is formed by uniting the three dimensions of "normal" space with time – producing "space-time". Objects such as galaxies, stars, planets, human beings, atoms, electrons and photons move within space-time in such ways that the distances along their tracks are minimized. Paths of this type are called geodesics. If space-time is curved in some way, then the geodesics are no longer straight lines but are also curved. A three-dimensional analogy of this is given by paths over the surface of the Earth. The shortest distance between two points on the surface follows the line of a great circle and is not the "straight" line obtained by moving along a constant azimuth.

The presence of a mass distorts the geometry of space-time in its region. General relativity thus replaces the idea that a mass exerts a gravitational force upon another mass with the concept that each mass distorts space-time in its own way. If the masses are moving with respect to each other, then the distortions of space-time lead to those motions appearing to us to be similar to motions under the "force" of gravity. The predictions of general relativity differ slightly from those of the Newtonian theory of a force of gravity – if this were not the case then there would be no advantage in using general relativity. One well-known difference is that the Newtonian predictions of the orbital motion of Mercury disagree slightly with its actual motion, whereas general relativity predicts the motion correctly.

Another difference between the two theories is in the paths that they suggest will be taken by light rays (photons) through gravitational fields. Using general relativity, the gravitational field of the Sun is calculated to deflect the paths of photons just skimming its surface by 1.75 seconds of arc and that is exactly the value that is observed. Newtonian physics however gives a deflection of only half that value – 0.875 seconds of arc. Larger and denser masses distort space-time to larger degrees than does the Sun leading to increasing deflections of light beams traveling in their vicinity. For a high enough mass and density the paths of all photons originating at its surface become so curved that they lead back to the surface. No form of e-m radiation, including light, can then leave such an object to emerge into the outer universe and we have a black hole.

The only properties of a black hole that are measurable by an outside observer are its mass, angular momentum (i.e. its spin) and its electric charge. If we set the latter two properties to zero for the moment then we get the simplest form of black hole, which is named after Karl Schwarzschild. Given the complexities that can be involved in general relativistic calculations, the size of a Schwarzschild black hole – called the Schwarzschild radius (R_s) – is found remarkably simply. For the Sun it is, to a good approximation, three kilometers. For black holes of other masses, the radius is then directly proportional to the mass (Table 5.1).

The event horizon is the name given to the envelope of points around a black hole from whence nothing can escape into the rest of the universe. For a Schwarzschild black hole this is simply a sphere whose radius equals the Schwarzschild radius.

Since most objects that we can observe in the universe are rotating its is unlikely that any black holes that may exist will be Schwarzschild black holes. Angular momentum is conserved during the collapse of a larger object to form a black hole and so the final black hole will be rotating very rapidly. Black holes that are rotating but not electrically

Table 5.1 The radii of Schwarzschild black holes.

Mass (kg)	Mass ($M_\odot$)	Radius (km)	Radius (AU)
2×10^{30}	1	3	0.00000002
2×10^{31}	10	30	0.0000002
2×10^{32}	100	300	0.000002
2×10^{33}	1,000	3,000	0.00002
2×10^{34}	10,000	30,000	0.0002
2×10^{35}	100,000	300,000	0.002
2×10^{36}	1,000,000	3,000,000	0.02
2×10^{37}	10,000,000	30,000,000	0.2
2×10^{38}	100,000,000	300,000,000	2
2×10^{39}	1,000,000,000	3,000,000,000	20
2×10^{40}	10,000,000,000	30,000,000,000	200

charged are called Kerr black holes. The event horizon of a Kerr black hole will be spherical and with a smaller radius than that of a Schwarzschild black hole with the same mass. However unless the rotational energy involved is truly enormous – a significant fraction of the rest mass energy[34] of the matter forming the black hole – then to a good approximation its size will still be given by the Schwarzschild radius. On the other hand, bodies that have a net electric charge are rare in the universe – stars and many gas clouds are so hot that their material is highly conductive and any electric charges that do build up are quickly discharged. We will not therefore concern ourselves further with electrically charged black holes.

Objects falling towards a black hole will normally possess some angular momentum with respect to the hole; they will therefore not fall directly into the hole but go into an orbit around it. This is of great practical significance as far as the model suggested for AGNs is concerned since in that model the AGN's energy is supplied from material

[34]The rest mass is the "actual" mass of a particle or other object, i.e. the mass measured by an observer who is at rest with respect to the object, so that there are no relativistic changes to the mass. The rest mass energy is then that mass converted to energy using $e = mc^2$; one kilogram converts to 9×10^{16} J (about 20 megatons of TNT).

falling into an accretion disk which is in orbit around the black hole, not from material falling into the hole itself (Sect. 5.2). Orbiting a black hole however is not just a simple extension of orbiting a planet or a star. In particular, stable orbits are only possible at some distance from the black hole. For a Schwarzschild black hole the radius of the innermost stable orbit is three times the Schwarzschild radius. For a Kerr black hole the innermost stable orbit has a smaller radius than that of a Schwarzschild black hole, and that radius decreases to a limiting value of one and a half Schwarzschild radii as the hole's angular momentum increases. Material inside the innermost stable orbit must have some radial motion and that will normally mean that it is falling into the black hole. The implication of this for modeling AGNs is that the accretion disk surrounding the central black hole will have an almost empty central region whose radius is that of the innermost stable orbit.

Outside their event horizons, Kerr black holes possess another "surface" known as the static limit. It is an elliptically-shaped envelope of those points where, in order to remain stationary with respect to the rest of the universe, material must move at the speed of light with respect to the black hole. The effect arises because the rotation of the black hole induces rotation in nearby material[35]. Inside the static limit, any object will be moving to some extent with the black hole's rotation as seen by an observer in the outside universe. The region between the static limit and the event horizon is termed the ergosphere because within it some of the rotational energy of the black hole may be extracted into the outside universe. This energy producing process is named for Sir Roger Penrose and is discussed in Sect. 5.1.3.

[35]This is known as the Lenz–Thirring effect and occurs for any rotating object, not just black holes. The magnitude of the effect is small, though, for non-black holes – one rotation per ten million years, for example, close to the Earth. Details of the effect are beyond the scope of this book and are left for interested readers to pursue for themselves (see Appendix 1).

5.1.2 Do Super-Massive Black Holes Really Exist?

Despite the widespread assumption that massive and super-massive black holes lie at the hearts of AGNs and perhaps of most galaxies, compelling evidence that the central mass is a black hole and not something else exists for only three galaxies – the Milky Way Galaxy, M 31 and M 106 (NGC 4258). The alternatives to black holes must be dark – 1% or less of the brightness per unit mass of the Sun. They must also be of high density – at least a million solar masses per cubic light year (thirty million solar masses per cubic parsec) – this compares with a density around the solar neighborhood of 0.002 solar masses per cubic light year (0.06 solar masses per cubic parsec) but is many orders of magnitude less than the density of a black hole. Possible black hole alternatives with these properties include clusters of stellar remnants such as white dwarfs and neutron stars or even stellar-mass black holes, clusters of brown dwarfs or of low-mass stars and concentrations of dark matter[36].

Proof that the central object is a massive black hole would ideally come from the observation of some phenomenon that could not occur for any of the black hole alternatives. One such observation might be that of relativistic velocities within a few Schwarzschild radii of the center of the nucleus. However this would require angular resolutions of microarcseconds for even the closest galaxies and is therefore way beyond the capabilities of any currently existing or envisaged instruments. Recently, the Advanced Satellite for Cosmology and Astrophys-

[36]The movements of objects within many galaxies suggest that they contain up to 10 times the amount of mass as that represented by the visible stars and gas clouds. The extra mass is called dark matter because it has yet to be detected directly. It nature is unknown – small black holes, brown dwarfs, neutrinos and exotic particles such as WIMPs (Weakly Interacting Massive Particles) have all been suggested, but no-one knows what the true nature of dark matter is.

ics (ASCA) has observed x-ray emission lines due to iron that have been broadened by velocities of up to 100,000 km/s ($0.3c$), but it is not clear whereabouts these velocities are occurring. In practice therefore, the process of "proving" the central object to be a black hole is one of showing that the outer limits of the volume containing the central mass are sufficiently small that the alternatives to black holes become much less likely explanations than black holes themselves.

Limits to the volumes containing the central masses have so far been obtained via direct observations and through observations of the movements of stars, of ionized gas clouds and of masers close to the centers of galactic nuclei. These have suggested some 50 galaxies wherein the central mass must either be a black hole or one of the alternatives. For the three galaxies mentioned above the size constraints have eliminated all possibilities except black holes or concentrations of dark matter.

The center of the Milky Way Galaxy is marked by the radio source Sgr A* (Fig. 5.1). It has been observed recently using the Very Long Baseline Array (VLBA). This is a VLBI that comprises 10 radio telescopes distributed over the USA and which can reach 0.1 mas resolution at the shortest wavelengths. The radius of the emission region is found to be no more than 1 AU (150,000,000 km) within which there is a mass of at least 0.04 megasuns and more likely one of 4 megasuns. The Schwarzschild radius of a 4 megasun black hole is 12,000,000 km so even this observation is not capable of resolving the central mass if it is a black hole. Nonetheless it rules out the possibility that the central mass could be any type of cluster of objects – over 40 million brown dwarfs would be needed, for example, and their average separations would then be less than a million kilometers which is only about five times their own diameters. The huge number of collisions that would result within such a crowded environment and the energy that would be released are not consistent with the weakness of the emissions from Sgr A*. Similar comments would apply to a cluster of white dwarfs or neutron stars, etc., the collisions would be fewer because the objects are smaller and of higher masses than brown dwarfs, but when they did

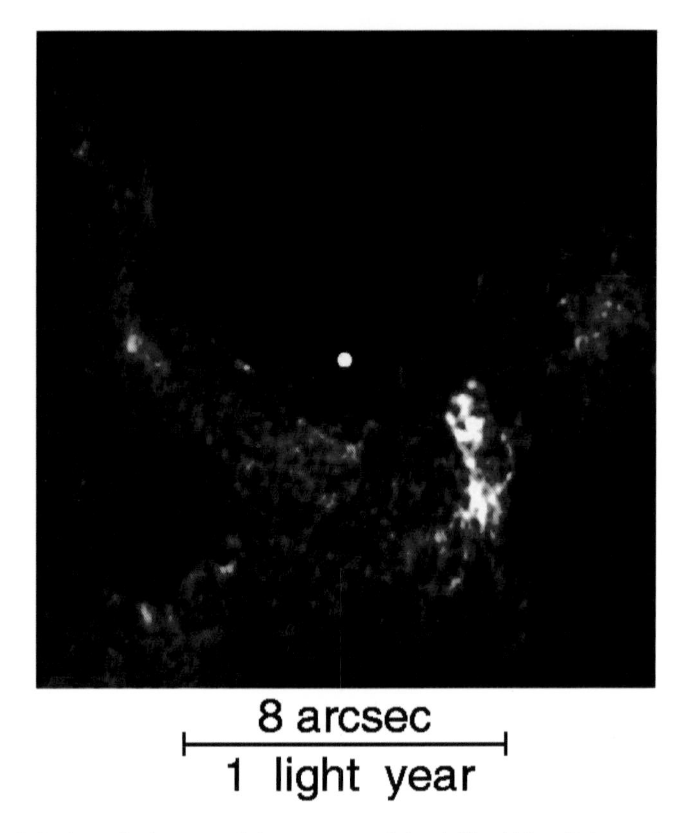

Figure 5.1 A radio image of the center of the Milky Way Galaxy obtained by the VLA. Sgr A* is the white dot at the center. (Image courtesy NRAO/AUI/NST, Jun-Hui Zhao, W.M. Gross and P. Smiley.)

occur they would be as spectacular as supernovae. Thus the only possibilities for the central mass of the Milky Way Galaxy is that it is a black hole with a mass of around four megasuns, or that it is composed of exotic particles such as (may) make up dark matter.

The Andromeda galaxy, M 31 (NGC 224), appears to possess two nuclei separated by about 0.5 seconds of arc, corresponding to a physical separation of about 6 ly (2 pc). In fact, recent HST observations have shown that the genuine (single) nucleus is surrounded by two structures that give rise to the apparent double nucleus. The first of these is an eccentric ring of red stars up to five light years from the center. The stars are in elliptical orbits and so pile up around the outer part of the ring where they are moving at their slowest speeds. This gives rise to the left-hand bright spot in Fig. 5.2. The right-hand bright spot surrounds the actual core of the galaxy and is a disk about one light year (0.3 pc) across that is composed of some 400 hot blue stars. The stars have orbital velocities of 1,000 km/s, implying a central mass of about 140 megasuns. As with the Milky Way Galaxy, such a mass inside such a volume combined with the quiescent nature of the nucleus is incompatible with that mass being anything other than a black hole or a concentration of dark matter.

Neither the Milky Way Galaxy nor M 31 are active galaxies, but M 106 (NGC 4258) is a Seyfert 1.9 galaxy. Water megamaser emission has been detected from its core by VLBI observations. The distribution of the masers suggests that they lie within a torus of material whose outer radius is 0.86 ly (0.26 pc), its inner radius is 0.4 ly (0.13 pc) and

Figure 5.2 An HST image of the "double nucleus" of M 31. The central black hole is hidden inside the bluish bright spot on the right. (Image courtesy of NASA, ESA and T. Lauer (NOAO/AURA/NST).)

whose thickness is less than 0.01 ly (0.003 pc) and which we view almost edge-on. The velocities of the masers at 900 km/s are consistent with their being in orbit around a 36 megasun central mass. Yet again, only a black hole or dark matter concentration can satisfactorily explain the observations.

In the three cases just examined the only adequate explanations consistent with the density of material at the core of the galaxies was either the presence of a black hole or of a concentration of dark matter. So how can we decide between these two possibilities? The answer at the moment is that we cannot. The nature of dark matter is not known, but its existence is inferred from the velocities of visible matter (stars, gas clouds, etc.) within galaxies. These observations suggest that there may be a great deal of dark matter – perhaps 10 times the amount that we see in the form of stars and gas clouds – but that its distribution is uniform and covers a region several times larger than the volume of the visible galaxy. There are no observations suggesting large concentrations of dark matter anywhere else within the galaxy or its halo. It has been suggested that dark matter could take the form of small black holes, neutron stars, brown dwarfs, etc., but we have already eliminated clusters of these objects as possible explanations for the central masses of the Milky Way Galaxy, M 31 and M 106. Thus we are left with the remaining suggestions for the nature of dark matter – neutrinos and WIMPS (Weakly Interacting Massive Particles). Neutrinos certainly exist and have been detected and studied coming from the Sun and the 1987 supernova in the Large Magellanic Cloud as well as from man-made nuclear reactors. Their masses however are far less than that of an electron and so given the slightest amount of kinetic energy, their speeds are close to that of light. It is thus difficult to envisage their remaining trapped inside a sub-light-year volume for the hundreds or thousands of millions of years required to explain AGN lifetimes. As for WIMPs – they have never been detected, they may or may not exist, and if they do then it must be in the form of exotic particles such as hypothesized companions to photons called photinos. Dark matter may of course be in some other form whose existence is currently completely unsuspected, in which case it could still make up the central masses of galaxies. For the moment, however, most astronomers simply take it for granted that the Milky Way Galaxy, M 31 and M 106 have massive black holes at their centers, and by extension that massive and super massive black holes are at the hearts of all AGNs and of many other types of galaxy.

The central masses – which we will now take to be black holes – can have their masses measured by determining the orbital motions of stars and gas clouds surrounding the core. Such observations, though, are difficult and lengthy and require the largest of instruments, so the number of galaxies and AGNs with accurately known central masses is still small. Even with the present comparatively few mass determinations though, it has become clear that the mass of the central black hole is related to that of the surrounding bulge of its host galaxy. The bulge in the case of an elliptical galaxy is the galaxy itself, for a spiral galaxy it is the galaxy's nucleus. The mass of the central black hole then turns out to be around 0.2% to 1% that of the bulge. Thus an estimate of the size of its black hole can be given for any galaxy through measurements of the size of its bulge. The small nearby Sc spiral galaxy, M 33, for example, has almost no central bulge and its central black hole has a mass of no more than 3,000 solar masses and it may not exist at all, while the Sombrero galaxy (M 104) with its huge bulge may have a central mass as large as 1,300 megasuns and the black hole in the giant elliptical galaxy, M 87, may be up to 3,000 megasuns.

5.1.3 How to Generate Energy Using a Black Hole

There is a widely held view that once something has fallen into a black hole it can never re-emerge and this forms the basis for many a dramatic episode in sci-fi stories. While it is true that the beautiful maiden abandoned by the villain to fall into a black hole (and of course rescued in the nick of time by the hero) cannot re-emerge as the same beautiful maiden, the energy represented by her mass (through $e = mc^2$) *can* get out again. So that black holes themselves are a type of energy source.

The process whereby this happens involves some subtle consequences of sub-atomic physics that are beyond the scope of this book. The interested reader is referred to sources in Appendix 1 for further information. The net result – as realized by Sir Stephen Hawking in 1975 – is that black holes are not truly black and they continuously radiate small amounts of energy. This Hawking radiation takes the form of thermal radiation (Box 2.5) and so its existence is equivalent to saying that black holes have a temperature above absolute zero. Unfortunately for anyone attempting to model AGNs the energy emitted by a massive or super-massive black hole is infinitesimal – for a 1,000 megasun black hole the temperature equivalent of its radiation is just 0.00000000000 000006 K (6×10^{-17} K) above absolute zero. Thus even though such black holes are radiating energy, they are also receiving vastly more from the 2.7 K microwave background radiation that pervades the whole universe plus starlight, cosmic rays, etc.

If Hawking radiation is no help in explaining the energy sources of AGNs, the Penrose process may hold out more hope. Named for Sir Roger Penrose who suggested its possibility in 1969, this provides a means of extracting rotational energy from a Kerr black hole. A particle within the ergosphere (Sect. 5.1.1) is required during this process to split into two. One of the halves then falls into the black hole in such a way that it reduces the black hole's angular momentum and rotational energy. The second half of the original particle is then flung out of the ergosphere carrying that excess energy with it. In this way up to 29% of the black hole's rest mass energy could be extracted[37], although the complexity of the required behaviors for the particles mean that it is unlikely to be the main source of AGNs' energies. A related process however, sometimes

[37]For comparison the conversion of hydrogen into helium, which provides the energy of the Sun and of most other stars, releases only 0.7% of the rest mass energy of the particles involved.

called the magnetohydrodynamic (MHD) Penrose process, is currently looking more promising and furthermore naturally results in twin opposed jets being flung away from the black hole (Chap. 4). An accretion disk around a black hole is likely to contain magnetic fields that have been intensified during the material's infall. As the magnetized plasma falls into the black hole it passes through the ergosphere where the magnetic fields become highly twisted. The magnetic fields then drive the plasma in two directions – inwards towards the black hole in such a way that the hole's rotation is opposed and outwards with increased energy into the two jets. Attractive though this model seems, it has yet to be fully worked out and integrated into a model for AGNs, furthermore, like the basic Penrose process, the physical requirements for its operation may make its occurrence in practice uncommon.

Thus, at the moment, we are left with no certain method of powering AGNs directly from central black holes. That is not the end of the story though – even if energy cannot be extracted from the black hole itself, it can be extracted from material during its descent towards the black hole. An object falling to the event horizon of a black hole from a large distance away will release potential energy equal to half its rest mass energy. If the object has an uninterrupted fall, then that energy will be converted to kinetic energy and the object will be traveling at close to the speed of light as it nears the black hole. Both the object's kinetic energy and its rest mass energy will then be absorbed by the black hole. However, if the fall is interrupted – perhaps by an accretion disk getting in the way – then that kinetic energy can be converted into other forms of energy and some or all of it may go on to power an AGN.

Imagine, then, a super-massive black hole surrounded by a supply of fuel, i.e. a dense cluster of stars, planets, gas clouds, etc. a few light years in size. The orbital motions around the black hole will lead to collisions, particularly amongst the physically large gas clouds. Random motions will tend to cancel out during such collisions and the system will settle down towards becoming a disk with an overall rotation around the black hole. It might be expected that the rotational plane of the disk would coincide with the black hole's equator, but this need not be the case. Super-massive central black holes probably come into being as a result of several mergers of smaller black holes and these could leave the final black hole with its rotational axis in any orientation. As the disk of gas becomes dense enough, planets and stars may become entrained within it as well. Even if this does not happen, stars or planets in orbits inclined to the disk will pass through it regularly, losing energy via friction as they do so and slowly spiralling inwards towards the black hole. During these collisions and other interactions, kinetic energy from orbital and radial motions is largely converted into the thermal energy of the gas particles, so heating up the disk.

Once the accretion disk is established it will not be stable. The gas clouds, stars and planets are in orbit around the black hole. The inner parts of the disk have shorter orbital periods than the outer parts and the viscosity of the gas tries to speed up the outer parts of the disk and to slow down the inner parts so that those periods become equal. Because the particles are in orbit however, such an equalization does not occur, instead the inner parts of the disk lose angular momentum, causing them to move into orbits closer to the black hole, which have even shorter periods. Conversely the outer parts of the disk gain angular momentum and move into more distant orbits with even longer periods. The transfer of angular momentum outwards via the viscous drag within the gas thus exacerbates the difference between the orbital periods of the inner and outer parts of the disk and the process continues. The net effect is that of driving the outer parts of the disk away from the hole and the inner parts in towards the hole. The potential energy released as material moves inwards is converted first into kinetic energy and then into heating the inner parts of the disk – perhaps to temperatures of 10^6 K or higher – which may then power other processes leading to the visible AGN. Once the infalling material descends below the position of innermost stable orbit, it will collapse rapidly into the black hole with little further energy being emitted into the outside world. In theory 50% of the rest mass energy may be released before the material falls into the black hole. In practice efficiencies of about a fifth of this are to be expected, i.e. a conversion of 10% of the rest mass energy into the visible phenomena of the AGN. For a bright quasar emitting 10^{40} W, this would imply an infall rate into the central black hole of around 10–20 solar masses of material per year. The accretion rate may need to be much larger than this estimate if the fuel is in the form of stars, not gas clouds. For black holes above a few hundred megasuns in mass stars will not be disrupted by tidal effects before reaching the event horizon. They will therefore fall in "whole" and contribute relatively little energy to the accretion disk. Smaller black holes will disrupt stars before they reach the event horizon and the resulting surge in mass accretion as the disrupted star interacts with the accretion disk may lead to outbursts in the activity of the AGN.

Now the estimate just made for the accretion rate needed to power an AGN does not depend upon the mass of the black hole; 10 solar masses falling into a 1 solar mass black hole via an accretion disk will release the same energy as 10 solar masses falling into a 1,000 megasun black hole via an accretion disk. So why do we need a super-massive black hole for AGNs? There are two parts to the answer to that question. Firstly, when we can measure the central masses of AGNs, they are *observed* to have megasun to thousands of megasun values. Secondly although a small black hole can in theory lead to the generation of as much energy as a massive one, the situation will be unstable. The reason for the instability is due to radiation pressure. In everyday life we are unaware that light (or any other type of e-m radiation) exerts a pressure since the brightnesses of commonplace light sources are low, but careful laboratory experiments confirm that it does exist. Radiation pressure increases as the intensity of the radiation increases, so that at the surface of the Sun, for example, radiation pressure amounts to about 0.2% of the gas pressure. Stars' luminosities intensify very rapidly as their masses rise and so for stars containing 20 solar masses the radiation and gas pressures are equal.

Now the stars that we observe in the sky are stable in the sense that the outward forces acting upon them due to pressure balance the inward forces due to gravity. Stability is ensured when gases provide the outward pressure, since should the star expand slightly for any reason, the gas pressure will fall below that required to balance gravity and the star will contract back to its previous size. Radiation pressure will also decrease if the star expands but it does so at the same rate as the magnitude of the gravitational forces decreases[38]. Thus the ratio between radiation pressure and gravitational forces remains constant when the star expands (or contracts). When stars' masses reach around 100 times that of the Sun, gas pressure becomes insignificant and gravity is balanced almost entirely by radiation pressure. In this situation, if for some reason (perhaps due to evolutionary changes in the star) the radiation pressure should exceed gravity, then the star will start to expand. Unlike the situation when gas pressure dominated though, now the radiation pressure remains in excess of gravity after the initial expansion and so the expansion continues – leading to the disruption and loss of at least the outer layers of the star and perhaps of the entire star[39]. Thus for a given luminosity, there is a lower limit (the Eddington limit) to the mass of the object if it is to be stable against radiation pressure. For the Sun, the Eddington luminosity is about 10^{31} W, that is, 25,000 times its current brightness. The Eddington luminosity is directly proportional to the mass, so that for a 10^{40} W quasar, the minimum mass for the core region (i.e. essentially for its black hole) is 1,000 megasuns. For a more average AGN luminosity of 10^{38} W, the minimum central mass is around 10 megasuns. If the AGN is radiating at less than the Eddington limit, then the mass involved will be even greater than these figures.

5.2 The Unified Model

The bewildering diversity of apparently different AGNs and other types of galaxies (Appendix 2 and earlier chapters) has arisen partly for historical reasons and partly because galaxies and AGNs do have an assort-

[38]The forces exerted by gravity and by radiation pressure both vary as the inverse square ($1/\partial^2$) of the distance from the central mass or light source.

[39]This is why the most massive stars that we have been able to discover are only a little over one hundred times the mass of the Sun.

ment of differing properties and behaviors. Historically, as new parts of the e-m spectrum became accessible to observers or as new and greatly improved instruments were developed within existing observational regimes, entities with new and different properties from anything previously known would be discovered. Often, to begin with, it was not even clear whether these new objects were local to the Sun, within our own Galaxy, or at enormous extragalactic distances. Certainly it would be many years before it was realized that in some cases these new objects were just different manifestations of things that were already known about – and by then the "new" entity would have received its own label and classification system, etc. Even when duplications such as N-type galaxies simply being distant Seyferts have been sorted out (and that ideal state has not yet been achieved), there remain many classes of extragalactic objects that at first sight seem to have little in common with each other. The credit for distilling some order out this chaos belongs to many scientists, but the numerous theories, ideas and proposals that abounded in the 1970s and 1980s were first brought together to form the beginnings of a coherent system linking many types of AGN in 1989 by Peter Barthel of the Kapteyn Astronomical Institute—producing what is now called the Unified Model of AGNs. Barthel's *Astrophysical Journal* paper identifies beaming as intrinsic to all quasars and that quasars and radio galaxies are morphologically and evolutionary similar objects whose differences can be explained if we see them at different angles to their beaming axes. He suggests that quasars have their beaming axes within 45° of the line of sight while radio galaxies are oriented at more than 45°. He goes on to speculate that QSO's may be powerful infrared galaxies, which are also viewed from a preferred direction.

The unified model has been considerably developed since Barthel's initial paper and today, while still not without some problems, provides an integrated explanation for the properties and behaviors of many types of AGN. In summary, the model suggests that there is a central black hole whose mass can range from a megasun to several thousand megasuns, which is surrounded by an accretion disk. Energy from infalling material heats the accretion disk and the resulting thermal radiation is observed in the blue and ultraviolet parts of an AGN's spectrum where it is often called the "Big Blue Bump". The disk is embedded in turn within a hot rarefied region emitting infrared synchrotron radiation and x-ray free–free radiation. Further out still is an absorbing torus of warm molecular gas. Jets formed from ionized gases (a plasma) are ejected at speeds ranging from a few times the local speed of sound to relativistic values along axes perpendicular to the plane of the accretion disk. The broad emission lines originate in numerous individual dense gas clouds that are typically 1 AU across and which are in more or less random orbits around the central core. The narrow lines also originate from small dense gas clouds that lie outside the molecular torus. Fuel, mostly in the form of gas, is supplied to the inner region from the rest of the host galaxy.

Under the unified model, AGNs of various types result from only two physical differences combined with differing viewing orientations. The first of the physical differences, which results in radio-loud and radio-quiet objects, is the possession or otherwise of a pair of relativistic jets. High-speed jets, perhaps formed when the central black hole is rotating very rapidly, result in strong radio emission, while lower speed jets, or none at all, lead to weak radio emissions. Seyfert 1 galaxies and QSOs then are weak or zero jet sources viewed from above the plane of the accretion disk, while Seyfert 2s are the same basic objects seen from a point of view that is closer to the plane of the accretion disk. Strong jet sources we see as blazars when viewing down the line of the jet, as quasars when we see them at a slightly greater angle, as a broad-line radio galaxy from a great angle still and as a narrow-line galaxy when we see them from around 45° to 90° to the track of the jets. The second physical difference is simply the energy available to the AGN, which almost certainly depends upon the amount of fuel that is available and results in fainter (eg. Seyfert 1 galaxies) and brighter (eg. QSOs) AGNs. It is also likely that the central black holes are more massive within the more luminous AGNs.

The main features of the unified model for AGNs are summarized in Figs 5.3 and 5.4. For the radio-quiet AGNs we start with a massive

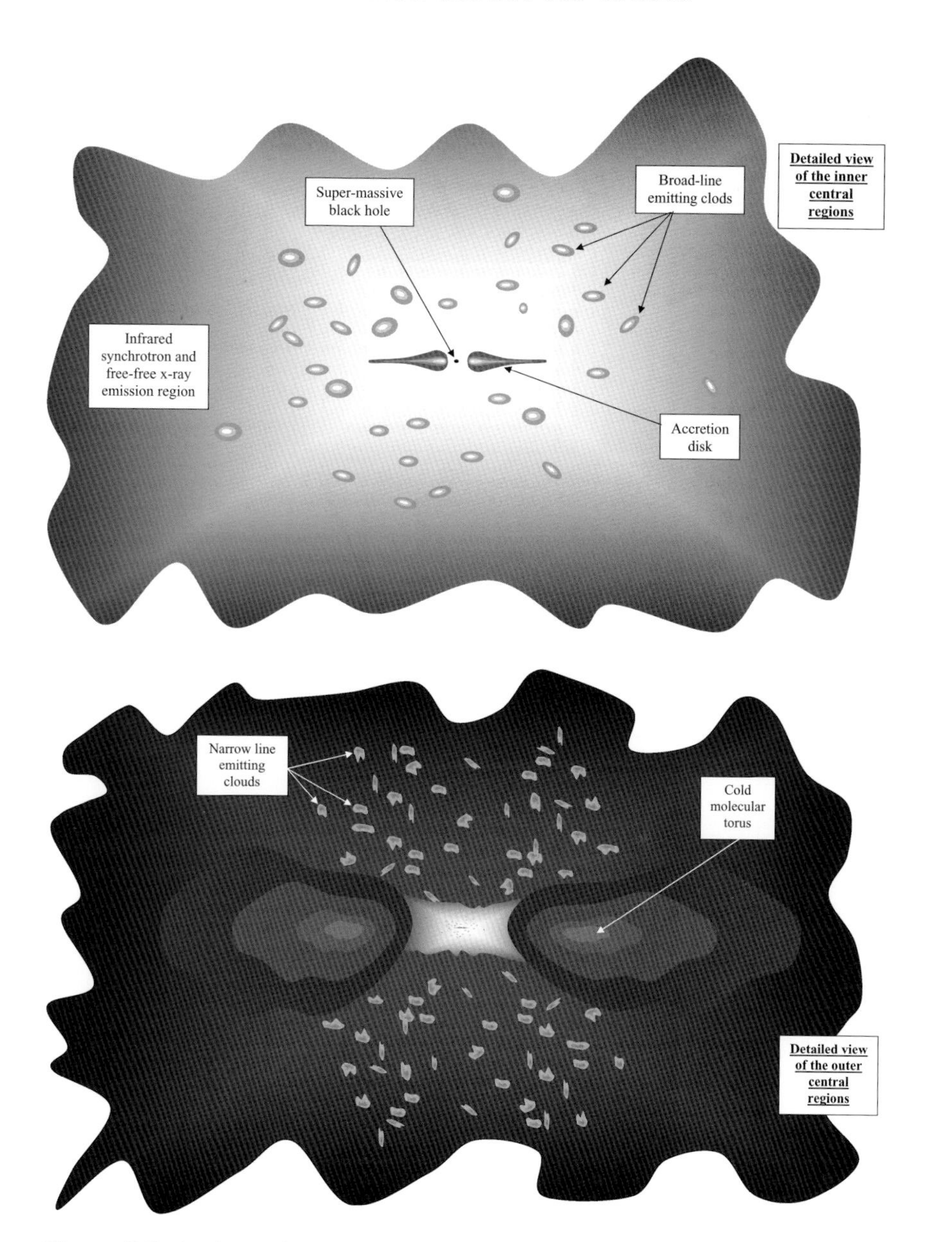

Figure 5.3 A schematic, not-to-scale, cross-section through the unified model for radio-quiet AGNs.

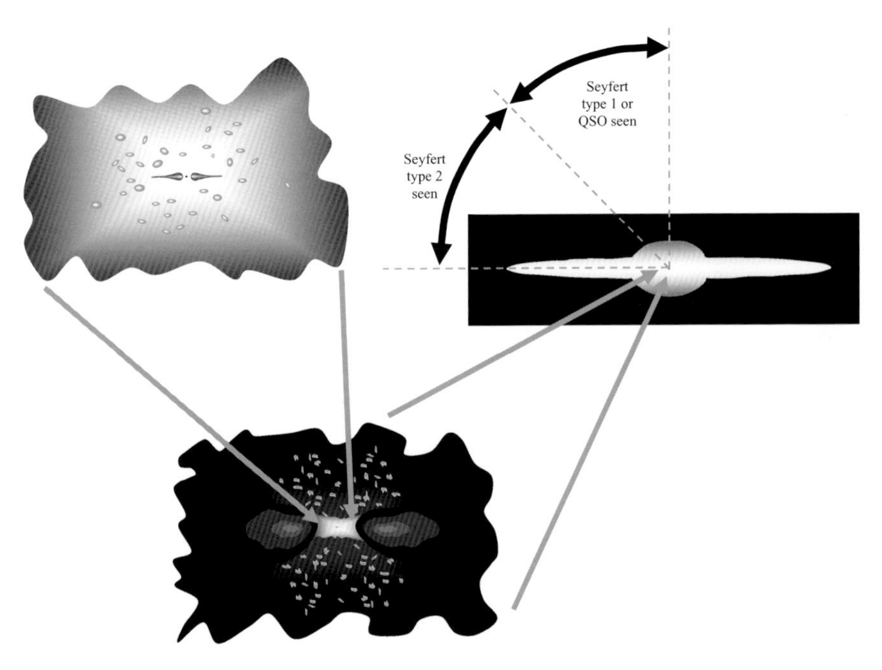

Figure 5.3 *Continued*

to super massive central black hole containing between a few hundred and a few thousand megasuns and whose size will probably be between about a tenth to ten billion kilometers (1–100 AU, 0.00001–0.001 ly, 5–500 μpc) across. The black hole will be at the center of a relatively clear region about three times its own size (see innermost stable orbit, above). That region will be permeated by intense and highly distorted electromagnetic fields and may have (smallish) amounts of material passing through it on the way to being absorbed by the black hole. Beyond the innermost stable orbit, the accretion disk will be found, perhaps extending out to a few tenths of a light year (~0.1 pc) from the center and with a thickness variously estimated to be between 0.001% and 10% of its diameter. There may be jets coming out perpendicular to the plane of the accretion disk (not shown in Fig. 5.3), but these are weak and slow compared with those of the radio-loud AGNs. The accretion disk in turn is embedded within an assemblage that may be several light years (1 to 2 pc) in diameter of small (a few AU) hot gas clouds. The gas clouds are orbiting around the black hole at high speeds and

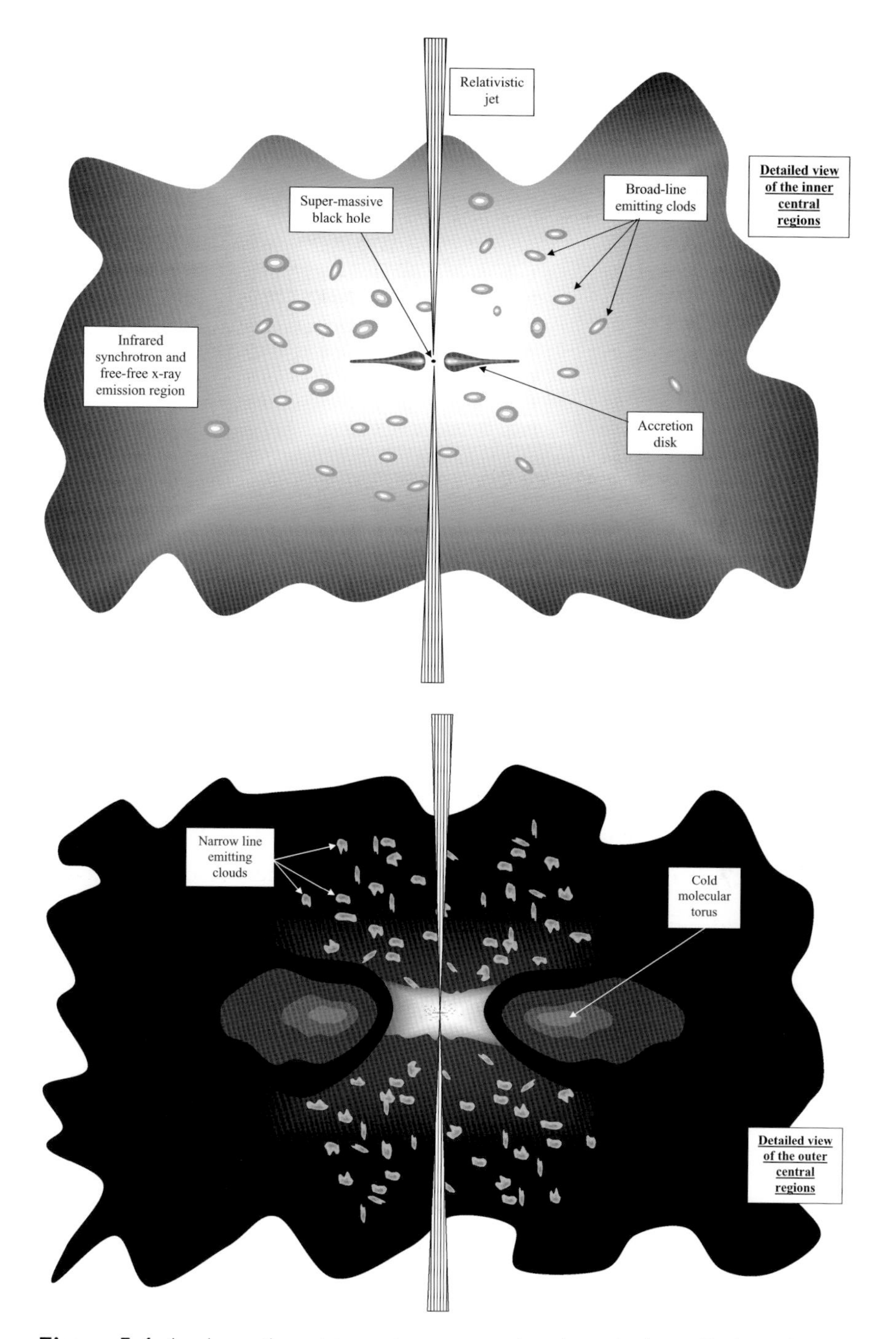

Figure 5.4 A schematic, not-to-scale, cross-section through the unified model for radio-loud AGNs.

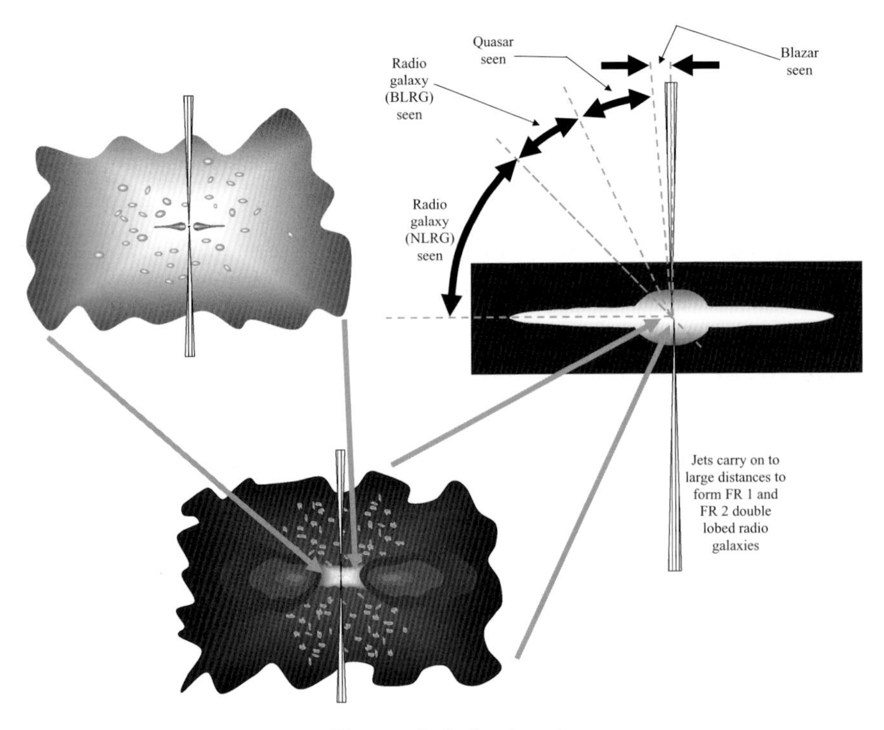

Figure 5.4 *Continued*

are emitting emission lines. The random nature of the clouds' motions means that the emission lines observed by a distant observer who cannot resolve the assemblage are very wide and it is this region that produces the broad emission lines observed in QSOs and Seyfert 1s. The clouds close to the hot accretion disk emit lines from highly ionized atoms, while those further out produce the lower ionization lines. It seems likely that the clouds dissipate quite quickly and so must be continually replaced by material from further out. Infrared synchrotron radiation and x-ray free–free or inverse Compton radiation are produced by the interactions of the electrons in the plasma and magnetic fields within this region.

Beyond the assemblage of broad-line-producing clouds lies a warm dense torus-shaped (ring doughnut) region of highly absorbing molecular gas and dust. The dust seems to be formed from silicate minerals and has been detected by its absorption at 10 μm and 18 μm in type-2 objects and by its emission at those same wavelengths in type-1 objects. The torus may be several hundred light years (100 pc) in its outer diameter and 100 or more ly (50 pc) thick. The inner radius of the torus is limited if its dust is not to be evaporated by the radiation from the central regions. The minimum inner radius for the torus of a 10^{38} W AGN is around half a light year (0.2 pc) while that for a 10^{40} W AGN is about 5 ly (2 pc). Many inner radii are probably larger than these minima, although NGC 4151 has recently had the inner diameter of its torus measured at 0.1 ly (0.04 pc) by reverberation mapping (Box 3.5). In addition to absorbing radiation coming from the core of the AGN, the gas and dust forming the torus re-emits that energy in the form of long-wavelength thermal infrared radiation. The torus is embedded within assemblage of line-emitting gas clouds. These clouds though, being further away from the black hole, have lower orbital speeds than those giving rise to the broad emission lines and so produce the narrow emission lines seen in most type of AGN. The central region of the FR 1 radio galaxy, NGC 4261, has been imaged by the HST and shows a dark absorbing disk some 400 ly (130 pc) across (Fig. 5.5). This structure may be the outer parts of the molecular torus or something separate from that, but either way it shows that disks can and do form in the centers of AGNs. Further out still (thousands to hundreds of thousands of ly, hundreds to tens of thousands of pc) we come to the relatively normal host galaxy which may, though, show signs of recent disturbance such as star burst regions or tidal distortions perhaps resulting from a recent encounter or merger with another galaxy. Also extending out to radii of 50,000 ly (20 kpc) or more there may, in some AGNs, be an extended narrow-emission-line region (ENLR). The lines from this region are very narrow (~50 km/s) and it is separate from the inner NLR. The ENLR can be quite asymmetric and it is thought to be due to interstellar medium gas ionized by ultraviolet radiation from the central AGN.

Now if we stand back and try to imagine how a distant observer would see a real object matching the unified model, then if that observer were to be positioned so that he/she can look into the central core of the AGN, then all the various emissions will be detected. Since the

Figure 5.5 Composite images of the FR I radio galaxy, NGC 4261. On the left is an optical image of the galaxy (white) combined with a radio image of the emission lobes (orange), which are nearly 90,000 ly (30 kpc) in length. On the right is an HST image of the galaxy's core. The luminous ring is composed of stars and surrounds a dark, cold absorbing disk at whose center a bright spot presumably marks the core of the AGN. (Images courtesy of NRAO, Caltech, Walter Jaffe/Leiden Observatory and Holland Ford/JHU/STScI/NASA.)

molecular torus absorbs most radiation very strongly, the need to see the core of the AGN requires the observer to be somewhere above the plane of the torus and within a cone whose half-angle to the perpendicular to the plane of the torus is in the region of 45° (Fig. 5.3). Such an observer will then find both broad and narrow emission lines in the spectrum coming from atoms in a range of levels of ionization, there will be intense thermal emission in the blue and ultraviolet regions from the hot accretion disk and there will be infrared synchrotron and thermal radiation and x-ray free–free radiation. These properties are exactly those of Seyfert 1 galaxies (Sect. 3.2.2) and QSOs (Sect. 3.2.4) providing that the energy supplies and, probably, the masses of the black holes are greater in the latter (remember from Sect. 5.2 that the Eddington limit implies higher minimum masses for the central regions as the total luminosity of the AGN increases).

If the observer's position lies at an angle greater than 45° or so to the perpendicular to the plane of the torus (Fig. 5.3), then the core of the AGN will be hidden behind the molecular torus. Since the torus is highly absorbing, radiation from the core will no longer reach the observer directly. He/she will then observe just the narrow emission lines and thermal infrared radiation from the torus. The emissions from the central region will still be brighter than those from the rest of the host galaxy though. These properties give us a Seyfert 2 type galaxy. As discussed in Sect. 3.2.4, whether or not type-2 QSOs exist is still problematical, but if they do then it seems likely that the unified model or some slight variation upon it will account for their properties.

Turning now to the radio-loud AGNs, the basic model is similar to that for the radio-quiet AGNs except for the addition of high speed, sometimes relativistic, jets of plasma being ejected from the region of the black hole and perpendicular to the plane of the accretion disk (Fig. 5.4). Explaining the origin of the jets (Chap. 4) is one of the major problems currently facing the unified model. From Figs 5.3 and 5.4 it can be seen that isotropically expanding material close to the black hole would be channelled into directions perpendicular to the plane of the accretion disk due to the obstacle to expansion along its plane formed by the accretion disk itself. However details of how such collimation would operate remain to be understood, and more importantly even the intense radiation pressure and electromagnetic fields in the inner region do not seem to provide sufficient energy to form relativistic jets. A possible clue to the explanation of jets' origins, but whose relevance is not yet completely clear, probably lies in the observation that radio-quiet AGNs are mostly associated with spiral galaxies, whilst radio-loud AGNs tend to reside within elliptical galaxies. Elliptical galaxies, especially the larger ones, are thought to originate from the collisions and mergers of smaller galaxies. The central black holes of the merging galaxies would also eventually be likely to collide and merge. Such collisions, though, are unlikely to be exactly head-on, but will be glancing ones that will spin up the new single black hole to very high rotation speeds. In some way not yet explained, rapidly rotating black holes may then produce high-speed jets, whilst those rotating more slowly do not. However if theory fails for the moment to explain why the jets are there, then we may still base the model upon the observations, which clearly show that they are indeed present.

Thus in Fig. 5.4 the radio-quiet unified model has been adapted by the addition of two high-speed jets (see Chap. 4 and Box 4.1 for an explanation of why two jets are present in the model even when only one is observed in many cases). The radio emission from radio-loud AGNs comes partly from within the jets themselves, but also from the jets' interactions with the surrounding gas. Although the basic model has altered little from the radio-quiet version, there are now four observing regimes giving different classes of AGN. As before, obscuration by the molecular torus leads to different AGN types, but we also have to take into account the effect of beamed emission from the jets (Box 4.1).

Just as previously, when the observer's position lies at an angle greater than 45° or so to the perpendicular to the plane of the torus (Fig. 5.4), then the core of the AGN is hidden. A distant observer will then see a narrow-line radio galaxy. Closer to the track of the jet, but before beaming becomes significant, the broad-line emission region will emerge from behind the molecular torus and we will observe a broad-line radio galaxy. Closer still to the track of the jet – perhaps 30° or less – and beaming starts to intensify the approaching jet and diminish the receding jet until only the former is detectable. At this angle of view, the object appears as a quasar. Finally when we look almost exactly down the track of the jet, the observed radiation comes almost exclusively from the jet and we see the featureless spectrum of a blazar. In all cases the jets are likely to continue well beyond the confines of their host galaxy and result in the double radio emission lobes of FR 1 and FR 2 radio galaxies.

As we have seen in earlier chapters, AGNs can vary in brightness over a range of timescales, with the most rapid changes often happening at the shortest wavelengths. Remembering that the physical size of the emission region imposes a minimum time for the occurrence of a significant change in luminosity (Box 3.4), the unified model suggests reasons for this. The x-ray emission can change in a few hours, but since this radiation originates from the central region, just outside the black hole, which has a physical size of a few AU, this is quite practicable. The whole of the region inside the gas and dust torus is no more than a few light years (pcs) across at most, and so changes over a few months to a year or two to any of the emissions from this region are again consistent with the model. Super-luminal motions (Chap. 4) are detected because in some AGNs jets material is ejected sporadically in the form of identifiable blobs. While the reason for this cannot be certain, it is explicable using the unified model via a feedback mechanism. Imagine the material falling in towards the accretion disk. It is unlikely that this material will be completely uniform, so on some occasions denser and more massive lumps may descend than at other times. It is easy to see that the impact of a larger-than-normal chunk of material might lead to higher-than-normal temperatures within the accretion disk. These in turn might lead to the ejection within the jets of denser and faster-than-normal blobs of material. Additionally though, pressure from the outburst within the accretion disk and jets might well slow down the still infalling material. There would thus be a hiatus in the delivery of material to the central regions, which would serve to separate the blob more clearly from the "normal" jet emissions. It is also possible that when the delayed infalling material does reach the accretion disk, it will do so "with a rush" – producing another outburst and another blob in the jet, thus neatly explaining those cases of AGN jets that are composed of several blobs.

5.3 Puzzles, Problems and Prospects

We have already encountered two of the major problems concerning the unified model for AGNs; whether the central mass really is a black hole or is some other form of concentrated mass – of which dark matter seems to be the only and remote possibility – and how jets can be driven to relativistic speeds. An explanation for the latter point will probably also solve the puzzle of why only 10% or so of the AGNs are radio-loud.

Other questions remaining to be answered include:

- How super-massive black holes could form within an aeon of the Big Bang (as required to form the oldest of the QSOs and quasars)?
- Why the abundances of elements heavier than hydrogen and helium in QSOs and quasars seem to be much the same whatever the redshift (i.e. age) of the object?
- Why the most luminous AGNs are associated with giant elliptical galaxies, while fainter ones tend have spiral galaxies as hosts?
- Why QSOs tend to have spirals or ellipticals as hosts, while the quasars are located within elliptical or disturbed galaxies?
- Do AGNs of one type evolve into another – and specifically – do QSOs and quasars originate as ULIRGs?
- What happens to AGNs when their activity dies away?
- Do type-2 QSOs and quasars exist?
- Why, even allowing for the effects of beaming, are so few counter-jets detected?
- How are jets collimated?
- How do the differences between the jets in FR 1 and FR 2 double-lobed radio galaxies arise?
- How can gas be driven in towards the black hole in sufficient quantities to power the AGN?
- Why are emission lines absent from BL Lacs' spectra?
- Are there hidden BLRs in all Seyfert 2s or are some of them genuinely without BLRs?
- What is the cause of the Baldwin effect?
- What is the power source in ULIRGs?

and

- How is star formation initiated within starburst galaxies?

Then, of course, there are Arp's proposals (Box 3.7), which discount the unified model entirely.

Nonetheless, at this stage we may justifiably conclude that the unified model provides a plausible explanation that links together many types, if not quite all, of AGN. The diversity amongst AGNs means that one can always find an AGN that fails to fit any selected aspect of the unified model. However, adapting a quotation from Sir Winston Churchill; "it's the worst theory we've got – except for all the others". The unified model does explain many aspects of AGNs in what seems to be a reasonable and physically sensible manner and if work remains to be done on it – well that's what will keep the next generation of galactic scientists in employment!

6
Origin and Evolution of AGNs

Summary

- Could the Milky Way Galaxy have been an AGN in the past?
- Could the Milky Way Galaxy still produce AGN-type phenomena?
- The Milky Way Galaxy/M 31 (possible) collision in three aeons from now.
- Life inside an AGN.
- The origins of super-massive black holes.
- The (possible) origins of AGNs of all types.
- The (possible) evolution of an AGN of one type into an AGN or galaxy of another type.
- The future development of AGNs.

6.1 The Milky Way Galaxy – Past, Present and Future

6.1.1 The Milky Way Galaxy

We have seen (Sect. 5.1.2) that at the center of the Milky Way Galaxy is the radio source Sgr A* and that this is almost certainly a black hole with a mass around four megasuns. Now in most AGNs where measurements have been possible, the central masses range from tens to thousands of megasuns. The Seyfert 2 galaxy, NGC 1068 (Fig. 3.4) for example has one of the smallest central masses, variously estimated at lying between eight and a hundred megasuns. The smallest known central mass for an AGN though is much less than this – for the dwarf Seyfert 1 galaxy, NGC 4395 it is calculated to be just 0.3 megasuns. Potentially then, the Milky Way has a central black hole massive enough for it to be an active galaxy. Although the Eddington luminosity (Sect. 5.1.3) for a four megasun black hole is 4×10^{37} W, which is just on the conventional borderline between Seyfert galaxies and QSOs and quasars, the Milky Way's black hole would probably only enable it to be a weak AGN at most. Sgr A* however is a faint radio source at just 10^{27} W. This is a million times fainter than the faintest galaxy recognized as an AGN, so clearly, despite its potential, the Milky Way is not an AGN at the moment.

If the Milky Way Galaxy is not currently an AGN, could it have been one in the past? Since there is a massive black hole at the center of the Galaxy, its current quiescence must be due to lack of fuel. There are several tens of megasuns of material within thirty light years (10 pc) of Sgr A* so there must be occasions when some of that material finds its way to the center and the black hole flares up. Data from ESA's Integral gamma-ray and x-ray spacecraft suggest the last such event might have been as recently as 350 years ago. The evidence for this comes from observations that show strong x-ray emission from a hydrogen gas cloud, known as Sgr B2, that lies 350 ly (100 pc) away from Sgr

A*. The radiation is thought to originate from gamma rays emitted from the region near Sgr A* that have been absorbed within the cloud and then re-emitted as x-rays. The gamma-ray luminosity of Sgr A* must thus have been far greater then than it is now. One estimate suggests that the intake of material into the black hole must have been around 10^{15} kg/s during the outburst. However with the typical 10% efficiency of conversion of mass to observable energy found for AGNs, this would only produce around 10^{31} W, still much fainter than the faintest AGN.

Further out from Sgr A* a number of large expanding shock fronts are detected at radio wavelengths. It seems likely that these have been driven by a powerful outburst from or near to Sgr A*. The rate of expansion suggests that the outburst occurred around fifteen million years ago and released some 3×10^{49} J over a million year period. While a hundred thousand supernovae could provide such an amount of energy, it seems more likely that the Milky Way Galaxy's central black hole was involved. If the analysis is correct, then the luminosity of the core of the Galaxy during the outburst would have reached 10^{36} W, i.e. a middlingly-bright Seyfert galaxy. The cause of the outburst is unknown although it could involve a cyclic process whereby the outburst blasts the central regions of the Galaxy clear of material for an interval and then material rushing inwards again produces another outburst.

Whether or not there was activity in the Milky Way Galaxy prior to this possible fifteen million year old outburst is unknown. Once its central black hole had grown to a megasun or more in mass, then any supply of fuel would have led to activity at some level. The activity however could only develop into full-blown AGN-type behavior if the fuel supply were to be large enough and continue for long enough to form the accretion disk, molecular torus and perhaps the jets. Smaller and briefer supplies of fuel would probably be more like a few supernovae going off in rapid succession or an x-ray flare within the galactic nucleus. We know that the Galaxy has collided with and cannibalized several smaller galaxies in the last few hundred million years, and it seems likely that such mergers were more frequent in the past when galaxies were closer together. Any such collision has the potential for forcing material towards the central region of the Galaxy and so of initiating an outburst. Thus while it is impossible to be definite, the balance of probabilities is that the Milky Way Galaxy *has* been active for one or more brief periods in the past, albeit almost certainly only as one of the weak AGN types.

What of the future? The Galaxy is likely to encounter and merge with more small galaxies over the next few aeons, but even within the local group of galaxies, the space motions of the galaxies are insufficiently well known for us to be able to predict which galaxies might be consumed and when. There is however one *big* exception to this – the Milky Way Galaxy and the Andromeda galaxy (M 31) are currently approaching each other at about 50 km/s. Given their separation of 2.4 million ly (740 kpc) and if they are truly on a collision course, they will come together in about three aeons from now at a closing speed in excess of 500 km/s. As we have discussed earlier (Sect. 2.3) the stars belonging to colliding galaxies will rarely, if ever, actually hit each other, but tidal forces will have major effects upon the structures of the galaxies. A computer model showing the possible results of a Milky Way Galaxy/M 31 collision are shown in Fig. 6.1. After about an aeon the two galaxies have merged and are settling down towards becoming a giant elliptical galaxy. Although the details of the interactions will depend critically upon the minutiae of the collision and will undoubtedly be different from those shown in Fig. 6.1, the overall result – a merger producing an elliptical galaxy – is probably correct.

M 31 has about twice the mass (700,000 megasuns) of the Milky Way Galaxy and its central black hole is about 140 megasuns in mass. The Milky Way Galaxy's four megasun black hole, even if it should combine with that from M 31, will thus not change things very much. However M 31's black hole is massive enough by itself to power an AGN provided that it is supplied with fuel. Whether it will be supplied with fuel during the collision is impossible to say, but the interactions are likely to be sufficiently complex (Fig. 6.1) that it would be surprising if some material were not to be channelled inwards. Thus we might

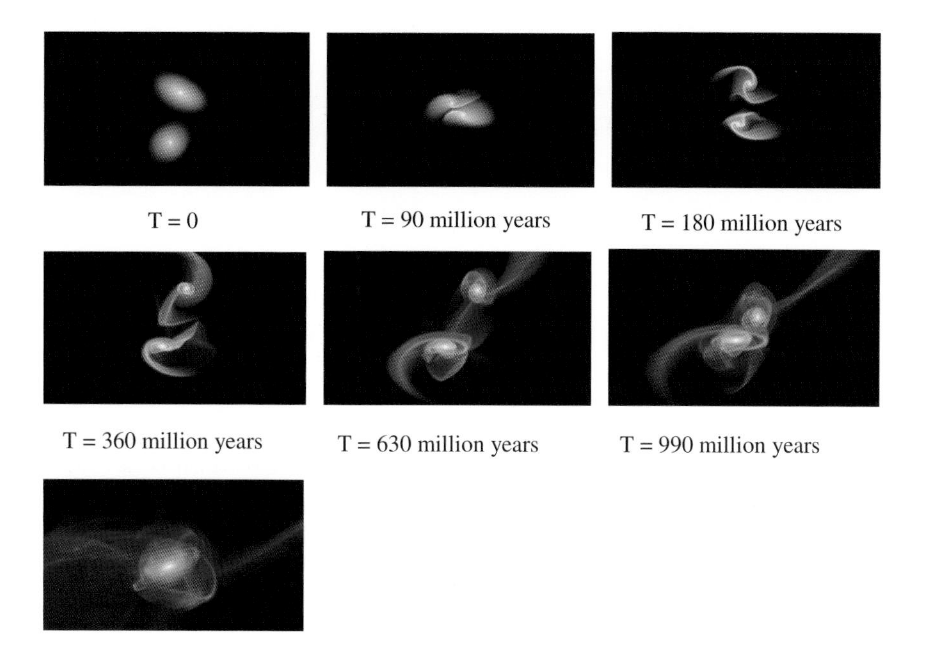

Figure 6.1 A computer simulation due to John Dubinsky of Toronto University of what may happen during the collision between the Milky Way Galaxy and M 31 in about three aeons from now. The Milky Way is the lower galaxy in the first image. The elapsed time from just before the collision is shown beneath each image in millions of years. The images are from a movie that may be seen in its entirety at http://www.cita.utoronto.ca/~dubinski/tflops/.

expect the merged galaxy to become an active galaxy for some or all of the time when the interaction is happening. It seems unlikely that a 140 megasun black hole could produce a QSO or quasar, but a fainter type of AGN is quite possible. In addition, of course, the interactions between the galaxies will lead to tremendous outbursts of star formation so that the merging object will be a starburst galaxy.

6.1.2 On Living Inside an AGN

In the unlikely event that human beings are still around and living within the solar system when the Milky Way Galaxy and M 31 collide, what will be their experiences? In the aeon or so before the collision, M 31 will dominate the night sky, appearing as a disk-shaped and probably brighter version of the Milky Way. In three to four aeons, the Sun will still be a main-sequence star although a little brighter than it is now. We may therefore find living on Mars, the asteroids or Jupiter's satellites more comfortable than the Earth. A direct collision between the Sun and another star is unlikely, but if it did occur then it would utterly destroy the solar system and any life that might then be in existence. More likely, although still improbable, is an encounter with another star that is sufficiently close for tidal forces to change the planets' orbits. Unless the tidal effects are very minor – and they could easily eject planets entirely from the solar system or send them crashing into the Sun – this is also likely to be fatal to life. Most probably, though, the solar system and our descendents will continue much as they had been doing before the collision. There would then be several possible fates for the Sun and solar system as a whole. If our descendents were really unlucky, they could find themselves part of the fuel for the AGN – and so truly find out what a black hole is like from personal experience. A second alternative could be for the Sun to be flung right out of the merging galaxy to become an isolated star in intergalactic space. The most likely possibility though, is that the Sun becomes a part of the merged galaxy but with a much more elliptical orbit than is now the case. The elliptical orbit will mean that the Sun will sweep through a large volume of the galaxy. Whether or not the merged galaxy has become an AGN, it will certainly be undergoing starbursts. The Sun and solar system will thus be quite likely to encounter supernovae during their travels. Unless the supernova is very close, the Sun and planets will be little affected by it, but the x-rays, gamma rays and cosmic rays from any supernovae within a few thousand light years (kpc) will endanger all living entities unable to shelter deep underground.

If the Milky Way Galaxy and M 31 do form a type of AGN some three or four aeons from now, the effects upon the solar system and any life forms within it could be either close to negligible or moderately serious. The difference depends upon the solar system's position within the galaxy. If the solar system is located so that intervening gas and dust clouds hide the central core (i.e. the position wherein a type-2 AGN would be seen by a distant observer, see Figs 5.3 and 5.4) then it will be protected from harmful emissions from the core. If the solar system were exposed to radiation from the core directly (i.e. a position from where a type-1 AGN would be seen), then the ultraviolet, x-ray and gamma-ray emissions could be dangerous, especially if the solar system were to be much closer to the core than is presently the case. The Earth's atmosphere, if still in existence and still containing oxygen, would provide protection for life at the surface of the Earth, but space travel would be very unsafe. Should the merged object develop relativistic jets one of which were to smash into the solar system, then it would again be the ultraviolet, x-ray and gamma-ray emissions which would cause the main problems, probably being intense enough to play havoc with the planets' atmospheres. The intense optical emissions of the beamed radiation would be seen as a star-like object whose apparent magnitude was between that of the full Moon and the Sun (depending on the solar system's distance from the core). Perhaps surprisingly, the relativistic particles forming the jet would be of relatively little consequence. The density within a jet is less than that of the interstellar medium (Sect. 4.3), so its effect would be equivalent to a small increase in the solar cosmic ray intensity.

6.2 Origin and Evolution[40] of AGNs

6.2.1 Introduction

This, of necessity, is a rather speculative section. We cannot in a human lifetime, or even several, watch evolutionary changes in any AGN, or indeed, in any other type of galaxy. We therefore have to rely upon

[40]Remember the previous comment (Sect. 1.2.1). Astronomers use the word "evolve" to mean the changes and development of a single entity over time (i.e. its life cycle).

inference (a better word would be guessing) to try and deduce how objects might change with time and whether one type of AGN might evolve into another. The old analogy, due to Sir William Herschel, of a botanist taking a 15-minute walk through a forest and then being asked to deduce the life cycles of the plants he/she had seen, which he used to illustrate how stars' lives might be understood, applies even more to galaxies and AGNs. The botanist would see trees, seeds, saplings, shrubs, seedlings, ground cover, decaying logs, etc. and might, given some background knowledge, deduce from this a realistic theory in which trees shed seeds which germinate into seedlings, then grow into saplings and eventually produce trees again, so re-starting the cycle. The scenario would remain a theory though – during the botanist's fifteen minutes of observation none of these changes would have been seen to occur. Moreover, it would be most unlikely that the botanist would be able to specify that this particular type of seedling would grow into that particular species of tree or that the well-decayed tree-trunk over there originally germinated from this type of seed, etc. The student of galaxies and AGNs is in exactly the position of Herschel's botanist – he/she must try to link the known properties of a large and disparate group of objects into a logical and hopefully, realistic, progression of types and events. The only real certainty in all of this is that change *does* occur – during the quasar era (Sect. 3.2.4) at redshifts around two to three (11 or 12 aeons ago) about one galaxy in a hundred was a QSO or a quasar, nowadays (redshifts zero to one, up to eight aeons ago) only one galaxy in a hundred thousand is a QSO or quasar – one thousandth of the quasar-era abundance. So the rate of production of QSOs and quasars has fallen drastically in "recent" times.

6.2.2 How Do Massive and Super-Massive Black Holes Form?

The unified model of AGNs (Chap. 5) tells us that massive and super-massive black holes underlie the behaviors of almost all types of AGNs.

The first necessity in developing a theory of AGN evolution is thus to understand how such black holes originate. One severe constraint on the process is that super-massive black holes must form very quickly by galactic time standards since the earliest known QSOs and quasars were in existence just a few hundred million years after the Big Bang, and the quasar era, when the vast majority of QSOs and quasars appeared, began only an aeon or two after the Big Bang.

One suggestion is that super-massive black holes are born directly from the collapse of super-massive stars. It is not clear how such stars would avoid fragmenting into thousands of smaller objects, nor why radiation pressure would not blow them apart when they exceeded their Eddington luminosities (Sect. 5.1.3), however if they do remain in one piece, then they would live their lives very rapidly indeed. In the unlikely event that the mass–luminosity relationship for normal stars holds for megasun stars, then their "main-sequence" lifetimes would be just three or four days. Once their energy sources were consumed such stars would collapse rapidly to become black holes since no other form of pressure exists that could support them against the inward pull of gravity.

Now while super-massive stars have the merit of forming black holes speedily, at least once they have themselves been created from collapsing gas and dust clouds, black hole formation via dense clusters of objects seems a more physically viable model. We have already seen (Sect. 5.1.2) that at the minimum density for the central object of the Milky Way galaxy (four megasuns within a volume of radius, 1 AU) if the material were to be in the form of numerous individual objects, then there would be frequent direct collisions between those objects whether they were brown dwarfs, stars, white dwarfs or neutron stars. During such collisions, objects would sometimes coalesce, and in the case of white dwarfs and neutron stars the merged objects would then be likely to exceed the mass limits for stable objects and so collapse to neutron stars or black holes in the case of white dwarfs or directly to black holes in the case of neutron stars. Brown dwarfs and normal stars might take a little longer to produce black holes since they would have to grow via collisions and mergers into massive stars and then evolve to white dwarfs and neutron stars, but the same situation would soon be reached. Once stellar-mass black holes had started to form, further collisions between the remaining non-black-hole objects and the black holes and between the stellar mass black holes themselves would cause the black holes to become more and more massive. Since the zone of influence within which it can capture material increases in size with the black hole's mass, the larger black holes in the cluster will tend to grow most rapidly. Eventually, most of the cluster's original mass will be contained within few largish black holes. For the purposes of driving an AGN a compact cluster of large black holes will probably be as effective as a single massive black hole. However perturbations by material surrounding the clusters and their own complex orbital motions are likely to reduce the numbers and increase the masses of the large black holes until just a single massive or super-massive black hole remains.

The model of massive black hole formation via collisions within dense clusters has one significant piece supporting observational evidence – the correlation between the masses of the central objects and those of the bulges of the host galaxies (Sect. 5.1.2). The black holes are found to have masses between 0.2% and 1% of the mass of the bulge – just the sort of connection that would be expected if the central cluster were simply the central region of the bulge.

Further growth of the central black holes can occur if material from the outer parts of the host galaxy finds its way down into the central regions – perhaps starting up the AGN as it does so. However after perhaps an aeon it must be expected that most of the available material will have been consumed. The black hole can then only continue to increase its mass via collisions between galaxies. Such collisions are likely to send fuel down towards the central black hole (Fig. 6.1), but also if both galaxies have central black holes, it can lead to a large increase in the black hole's mass as those two black holes collide and merge. Whether all this can occur rapidly enough for the first QSOs and quasars to start radiating just a fraction of an aeon after the Big Bang remains unclear, however as such QSOs and quasars *are* observed to exist, one of these processes for producing massive black holes that have just been outlined – or something else – *has* to do so.

6.2.3 Active Galaxies – Past, Present and Future

One of the criteria for identifying a starburst galaxy (Chap. 2) is that its rate of star formation greatly exceeds the rate sustainable over the galaxy's lifetime. The starburst event must thus be a relatively transient episode and starburst galaxies must have been some other type of galaxy before their starburst event and will go on to revert to some other type of galaxy after the starburst is over. Although starburst regions can occur within some AGNs, starburst galaxies generally do not have active cores. Instead, most show signs of undergoing or of having recently undergone an encounter with another galaxy, and it is the disturbance to the gas and dust clouds arising from the tides induced during that encounter that lead to the outbursts of star formation. Starburst galaxies are found at all distances – ranging from the nearby M 82 at 11 Mly (3.3 Mpc) away from us to 12,000 Mly (3,500 Mpc) where perhaps one galaxy in eight contains a powerful starburst. Starburst galaxies are thus a phenomenon that can occur today as well as throughout most of the life of the universe. This fits in well with the idea that starbursts originate through encounters, collisions and mergers between galaxies since such events can happen at any time. The higher proportion of starburst galaxies early in the life of the universe than today also fits an origin through collisions since the smaller volume of the universe in the past would have made collisions more frequent.

The duration of a starburst will depend upon the amount of gas and dust available to create new stars. Today that is rarely greater than 15% of the mass of the galaxy (Chap. 1), although the proportion must have been higher in the distant past since "normal" star formation is continually converting gas and dust into stars. Star-formation rates in starburst galaxies can be up to several thousand solar masses per year. If we take a medium-sized galaxy and assume that about a quarter of the available gas and dust inside it becomes compressed sufficiently to initiate star formation, then we find that a star-formation rate of a thousand solar masses per year is sustainable for only a few tens of millions of years. The highly luminous hot and massive stars that delineate the star-forming regions exist for only a few tens to a few hundreds of millions of years. Thus the lifetime of a starburst galaxy is probably limited to a few hundred million years. Before the starburst the galaxy could have been any variety of galaxy. After the outburst, the galaxy is likely present a highly disturbed appearance for some time before eventually settling down towards an elliptical form.

Spiral galaxies contain the highest proportions of interstellar material, so interactions involving two large spiral galaxies are likely to produce the most violent starbursts. The most luminous of the starburst galaxies – the ULIRGs – probably arise through spiral galaxy collisions. However in these galaxies, the starburst regions are not usually seen directly, but their existence is inferred as the sources of the energy that heats dust particles. The dust particles in their turn emit far-infrared radiation, which is what is then actually observed. However since the starbursts are not observed directly, other energy sources could take their place. This has led to the suggestion that young QSOs or quasars power some ULIRGs. Indeed, infrared spectroscopic observations from ESO's VLT imply that two-thirds of ULIRGs contain AGNs. It is thus possible that ULIRGs represent the long-sought type-2 QSOs and quasars (Sect. 3.2.4). Additionally or alternatively, the QSO or quasar may evaporate the outer regions of the ULIRG quickly and so emerge to be seen as a type-1 object. Whether or not all, or indeed any, QSOs and quasars originate from ULIRGs (plus perhaps the lower luminosity far-infrared galaxies – FIRGs) remains however to be proven.

In addition to the starburst galaxies where there has clearly been some recent or continuing interaction with another galaxy, some can be found that show no signs of any such an encounter. It is possible in these cases that tidal effects from a central bar are the trigger for their outbursts, but there remain others where there is no clear explanation for the origin of their enhanced rates of star formation.

Origin and Evolution of AGNs

The origin of Seyfert galaxies is currently a puzzle. Most likely they are just normal galaxies, probably spirals, which over time have amassed central black holes large enough to support AGNs. However it has also been suggested since the brighter Seyferts overlap in their properties with those of the fainter QSOs and quasars (Chap. 3), that Seyferts are "dying" QSOs and quasars. On balance this seems unlikely to be the case for the majority of Seyferts, although odd examples of this happening cannot be ruled out. The reasons for the doubt is firstly that there seem to be too many Seyferts for them all to have evolved from QSOs and quasars and secondly that the masses of the central black holes tend to be lower in Seyfert galaxies than they are in QSOs and quasars, suggesting that the difference is not simply one of age. A similar argument suggests that LINERs also originate as separate objects rather than being the last stages of Seyfert activity.

The future for AGNs of all types is the same – their activity must eventually fade away to nothing. We may easily see this given the rate of consumption of fuel by the central black hole that is required to power AGNs. This is estimated (Sect. 5.1.2) to be 20 solar masses per year for the brightest QSOs and quasars. With a large host galaxy (say 500,000 megasuns) and assuming that 1% of the host galaxy's mass is available to power the AGN, then the maximum lifetime for such a QSO or quasar is around a quarter of an aeon. This is a very rough estimate of the duration of most QSOs' and quasars' activities since it will be affected by a number of factors:

- a typical QSO or quasar has only 1% to 10% of the luminosity of the brightest examples;
- most host galaxies are probably of significantly lower masses than 500,000 megasuns;
- the duration depends upon the efficiency of conversion of fuel into energy of the AGN – this has been estimated at 10% (Sect. 5.1.3) but could be significantly higher or lower than this figure;
- the availability of fuel has been guessed at 1% of the mass of the host galaxy, but it could just as easily be 5% or 0.2%, and probably varies

from one AGN to another depending upon the host galaxy's circumstances.

Nonetheless the estimate of a fraction of an aeon cannot be too far out since the duration of the quasar era of about one aeon (from 11 to 12 aeons ago) limits the lives of QSOs and quasars to less than a single aeon. If the lifetime were to be longer than this then those QSOs and quasars formed 12 aeons ago would still be radiating 11 aeons ago and so the quasar era would not then be coming to an end. The lower luminosity AGNs could have lives significantly longer than one aeon – perhaps up to the age of the universe or longer, however since they are also consuming the material of their host galaxies, there must come a point when that material has been used up and the AGN ceases its activity.

Thus most QSOs and quasars have already come to the ends of their lives and we are left with super-massive black holes at the centers of the original host galaxies. The same must apply to many fainter AGNs and will apply to all in due course, although their central black holes are likely to be less massive than those of the QSOs and quasars. We have also seen that many normal galaxies already have large, if not massive, central black holes. Thus predicting the future of AGNs and most other types of galaxy reduces to answering the question of what will happen to a system of stars and gas and dust surrounding a central black hole as time passes?

The answer is that over the next few aeons we may expect to see occasional flare-ups and the return to some form of activity as individual stars, star clusters and gas clouds happen to have their orbits perturbed into the core of the galaxy (Sect. 6.1.1), so encountering the black hole. On longer timescales – perhaps ten thousand times the present age of the universe – the stars will end their lives and the galaxies fade away leaving the black remnants of stars and planets still orbiting the central black hole. It will take a thousand times longer still, 10^{17}–10^{18} years, for significant changes to occur within these galactic corpses. Over that interval the remaining "stars" will have undergone many close passages (though few direct collisions) with each other. The stars' orbits will be changed during such encounters – some perhaps gaining sufficient energy to leave the galaxy entirely and wander through space as independent bodies – others losing energy and so moving into orbits that take them closer to the black hole. Inevitably some of the latter will fall into the black hole. The net effect is for some stars to be lost entirely from the galaxy and for the remainder to huddle ever closer to the black hole until they are captured by it. If any stars manage to avoid either of these fates and remain in orbit around the black hole, then gravitational radiation will cause them to join their colleagues inside the black hole by the time that the universe is approaching 10^{20} years old. The future for AGNs and pretty well all other macroscopic objects within the universe thus lies either within a super-massive black hole or as an independent and incredibly isolated stellar cinder.

7 Active Galaxies Through the Telescope

Summary

- How to observe AGNs visually.
- Observing details for the brightest AGNs.
- How to image AGNs with CCD cameras.
- Boxes
  Dark adaptation and averted vision.

7.1 Visual Observations

Many observers probably have the impression that studying active galaxies requires vast observatories and humongous telescopes, because the majority of images that they see of active galaxies in books and astro-magazines are indeed those coming from such instruments. However, although having a 10-m telescope in your back garden would undoubtedly be useful, it is *not* necessary for observing active galaxies – indeed several can be seen through binoculars and a 0.2-m (8-inch) telescope will reveal enough of them to keep you occupied for several years. If you have an astronomical CCD camera in addition to a modest telescope, then you can obtain images that professional astronomers using major telescopes two or three decades ago would have envied.

Observing many active galaxies is actually easier than observing ordinary galaxies. QSOs, quasars, blazars and many Seyfert galaxies emit most of their energy from the central star-like AGN instead of spreading it out over the area covered by the galaxy. Thus if you can observe a star of magnitude 13^{m} you can probably observe a QSO, quasar, blazar or Seyfert galaxy of magnitude 13^{m} (and so on). A reasonably acute observer from a reasonably good observing site can see 6^{m} stars with the unaided eye. Using a telescope, the limiting visual magnitudes are listed in Table 7.1.

Practiced observers working from excellent observing sites, without a moon in the sky and using averted vision with fully

Table 7.1 Normal limiting magnitude for star-like objects observed visually through a telescope.

Telescope diameter (m)	Telescope diameter (inches)	Limiting visual magnitude
0.05	2	9.5
0.10	4	11.0
0.15	6	11.9
0.20	8	12.5
0.25	10	13.0
0.30	12	13.4
0.35	14	13.7
0.40	16	14.0
0.45	18	14.3
0.50	20	14.5

dark-adapted eyes (Box 7.1), can perhaps better these figures by one magnitude. Thus the brightest quasar (3C273, 12.86^m) with luck could be seen using a 0.2-m (8-inch) telescope, would have a good chance of being seen through a 0.25-m (10-inch) telescope and should be seen even from poorer sites through 0.3-m (12-inch) or 0.35-m (14-inch) telescopes.

Box 7.1 Dark Adaptation, Averted Vision and Other Odd Phenomena Involved with Eyesight

The "techniques" of dark adaptation (or night sight) and averted vision are essential to the successful observation of faint nebulous objects and anyone intending to study active galaxies must be familiar with their effects.

Dark adaptation will probably already be known to most people – when you go outside a brightly lit room on a dark night, you can see little of your surroundings for the first few seconds. After a couple of minutes you are likely to be able to see sufficiently well to be able to move around confidently and after half an hour your sight will have improved to the point that you can see perhaps 25% as well as in daylight (although not in color). The phenomenon arises because under high levels of illumination, light reception within the eye is mostly due to the cone cells in the retina, which produce color images[41]. The rod cells, which provide monochromatic vision, are saturated by bright lights and then work at very low efficiencies. When light levels are low, the chemicals within the rod cells slowly regenerate, taking 20–30 minutes to recover completely. As the rod cells regenerate, so the efficiency of their light detection improves, peaking at around 100 times that of the cone cells. When this improved response is combined with the increase in the size of the pupil of

[41] For a fuller description of the processes lying behind vision see, for example, the author's *Astrophysical Techniques*, 4th edn, published by IoP Press, 2003.

the eye from around 2 mm at high light levels to 7 mm at low light levels, the sensitivity of the eye is improved by a factor of a thousand or so.

Dark adaptation is thus just the process of allowing the eye's rod cells to regenerate their sensitivity. Any exposure to bright light will halt or reverse this process, so in order to become fully dark-adapted, the observer must remain under low levels of illumination for around half an hour. His or her vision will then have become vastly more sensitive and objects that earlier had been quite invisible through the telescope will now be seen easily. Observers trying to push the limits of what can be detected through their telescopes will need fully dark-adapted eyes.

Dark adaptation can be destroyed very quickly – a few seconds exposure to the light from a torch, for example, will ruin half an hour's waiting in the dark. Many astronomers use red filters on their torches or other lights since the sensitivity of the rod cells cuts off at long wavelengths and so their night sight should be unaffected. However most such filters allow too much orange light through and dark adaptation is still compromised by them. To be effective the red filter must transmit nothing with a wavelength shorter than about 600 nm – and this will be seen to be a very deep red indeed. The Wratten #29 filter is suitable for this purpose. If possible, once dark adaptation has been achieved the observer should use no lighting at all – records of the observations being fed into a mini-tape recorder for later transcription. Even with all these precautions, dark adaptation can be reduced if you see a bright object through the telescope. Since it is common practice for many observers to star-hop from easily found stars to the object that they want, they may need to wait for a few minutes, once in the right spot, for their night sight to recover completely.

Averted vision is a phenomenon that is also related to the rod and cone cells within the retina. These are not distributed at random, but the cone cells are concentrated into the part of the retina that normally receives the center of the eye's field of view. The rod cells are more plentiful around the parts of the retina involved with peripheral vision. Thus when you look directly at an object, the light falls mostly onto cone cells. However, if you look to one side of the object of interest then its light will fall onto a region of the retina that is richer in rod cells. Since at low light levels the rod cells have 100 times the sensitivity of the cone cells, the object may then become visible even though it could not be discerned when looked at directly. Averted vision thus requires the observer *not* to look directly at the object, but instead slightly to its side. It is a trick that is a little difficult to acquire, since faint objects will flash into view as the observer looks away from them – and one's instinct then is immediately to look at them directly again. However a little practice will soon enable the student to observe objects by not looking straight at them.

Some observers claim that their vision sensitivity is improved by hyperventilating (taking several deep breaths in rapid succession). If the effect is real, then it is of much less significance than dark adaptation and averted vision, but the reader may care to experiment for him/herself – after all, every little helps.

Finally it should be mentioned that the materials forming the eyes become less transparent with age. The retinas of a 60-year old person receive only about a third of the light that is seen by a 30-year old. Eye diseases can of course make this deterioration far worse. On the other hand if you have had a cataract operation, you may be able to see further into the ultraviolet than normal and since many AGNs brighten towards shorter wavelengths (the Big Blue Bump), they may become easier to see.

AGNs that are not effectively star-like sources are observable visually using the same methods as those that are used for ordinary galaxies. You will need the best observing site that is available (i.e. one with as little light pollution as possible). Of course, with a permanently mounted telescope, the observing site is fixed, but with portable instruments it may be possible to move to better positions. A light pollution rejection

(LPR) filter[42] may improve the contrast between the AGN and the background radiation, although it does also reduce the total intensity of the object as well. It is best to observe when there is no moon in the sky, although up to a half moon may be tolerated provided it is well away from the part of the sky in which you are observing. Generally speaking, the larger the telescope that you have available, the better. Low magnifications will probably give the best results since they provide the best contrast. Good dark adaptation is almost certain to be essential unless you are looking at the brightest of AGNs – and even then will help with making out their faint outer regions. Averted vision will be needed when trying for objects at the limit of the instrument.

Table 7.2 lists the brighter AGNs together with information to help the observer. The selection has been made on the basis of star-like objects having an apparent integrated magnitude (column 6) brighter than 16^m in reference 2 (see footnote), the optical counterparts of radio galaxies having an apparent integrated magnitude brighter than 14^m in reference 2, since these are likely to be extended but small sources and of other AGNs having a core magnitude (column 7) brighter than 12^m in reference 1, since these are likely to be larger extended sources. A few other objects such as 3C48 have been added for historical or other reasons. For star-like AGNs the apparent magnitude of the central 16 arcseconds of the AGN (column 7) is probably the best guide to their visibility. For extended objects, the integrated magnitude is listed in column 6, however the object's visibility will also depend upon its angular size (column 8), so column 9 lists a visibility class that is related to the object's brightness per unit area. Finally column 10 provides

[42]These filters block the main emission lines emitted by terrestrial sodium and mercury lights. They may also be called deep sky filters. If the AGN is has strong emission lines then an LPR filter may improve its visibility if these lines do not coincide with the regions of the spectrum blocked by the filter. However if the AGN's redshift moves its emission lines into such blocked regions, then it will become far harder to see. A deep red filter such as the Wratten #29, can reduce light pollution on CCD images since such detectors have a peak sensitivity in the near infrared (Sect. 7.2).

comments from observers using small to medium-sized instruments about the object, where these are available.

Table 7.2 Observing details of the brightest AGNs[43].

Column 1 – Names
- 3C – Third Cambridge radio catalogue, 1962
- 4C – Fourth Cambridge radio catalogue, 1965–67
- AO – Arecibo Occultation catalogue, 1967–70
- B2 – Second Bologna survey
- B3 – Third Bologna survey
- C – Caldwell catalogue
- CSO – Case stellar objects survey
- IC – Index catalogue (Supplements to NGC)
- I Zw – Catalogue of selected compact galaxies and of post-eruptive galaxies, 1971, F. Zwicky
- KAZ – New galaxies with ultraviolet excess, 1979, M.A. Kazarian
- M – Messier catalogue
- Mrk – Markarian galaxy
- NGC Catalogue number in Dreyer's *New General Catalogue of Nebulae and Clusters of Stars.*
- NRAO – National Radio Astronomy Observatory 1400 MHz radio survey, 1966
- OA, OB, OC, . . . OZ – Ohio state university radio source catalogue, 1967–74
- PG – Palomar–Green list of bright QSOs, 1983
- PKS – Parkes radio survey, 1969–75
- S2 – Strong source list number 2, 1972, H. Kuehr, et al.
- S4 – Strong source list number 4, 1978, I.I.K. Pauliny-Toth et al.
- SBS – Second Byurakan Survey
- Ton – Tonantzintla catalogue of blue stellar objects, 1957–59

Column 2 – The type of AGN
- BLL – Blazar – BL Lac type
- HPQ – Blazar – HPQ type
- LNR – LINER

[43]Data sourced primarily from:

Reference 1 – CDS catalogue VII/248 – *Quasars and Active Galactic Nuclei* (12th edn), M.P. *Veron-Cetty, P. Veron, 2006* (available at http://cdsweb.u-strasbg.fr/cats/VII.htx).

Reference 2 – *Sky Catalogue 2000.0*, Volume 2 – *Double Stars, Variable Stars and Non-Stellar Objects*, A. Hirshfield, R.W. Sinnott, Cambridge University Press, 1985, ISBN 0-521-27721-3.

Photo-Guide to the Constellations, C.R. Kitchin, Springer 1998, ISBN 3-540-76203-5.

Table 7.2 *Continued*

Q – Quasar
QSO – QSO
R – Radio source
RS – Radio source – spiral galaxy
RE – Radio source – elliptical galaxy
RD – Double radio source
RDS – Double radio source – spiral galaxy
RDE – Double radio source – elliptical galaxy
STB – Starburst galaxy
Sy – Seyfert galaxy. The subclass, if known, is denoted by one of the numbers; 1, 1.2, 1.5, 1.8, 1.9, 2.

Column 3 – Right Ascension for the epoch 2000.0
Column 4 – Declination for the epoch 2000.0
Column 5 – Constellation within which the AGN may be found
Column 6 – The integrated apparent magnitude of the AGN and host galaxy (when visible) in the yellow–green (V) part of the spectrum. The values quoted here are mostly from reference 2 (see footnote). This is only a partial guide to the visibility of extended objects and must to be combined with the size of the object to give the surface brightness. Often the bright nucleus of an AGN may be found as a star-like object when the rest of the galaxy remains invisible (see visibility class and observing notes, below).
(B) – indicates a blue magnitude, not visual.
v – indicates a variable magnitude.
Column 7 – The apparent magnitude of the central 16 arcseconds of the AGN in the yellow–green (V) part of the spectrum. This is may usually be taken to be a small or star-like object, and so if you can see/image a star of this brightness you should probably be able to do the same for the AGN. The values quoted here are from reference 1.
Column 8 – Angular size in arcseconds. Values are mainly taken from reference 2. NB the perceived angular size is likely to be quite different from these values and will depend upon the observer's visual acuity, degree of dark adaptation, the size and quality of the telescope, the quality of the observing site, especially the amount of background light, etc.
Column 9 – Visibility class – An estimate of the likely ease of observation of the object based upon its integrated luminosity and area. Extended objects are assumed to emit most of their energy within the central 10% of their radius and point sources are assumed to be broadened by atmospheric seeing to two seconds of arc. The classes range from A (greatest luminosity per unit area) to D (least luminosity per unit area).
Column 10 – Other comments about the object. Observing notes for visual observers (ON) are mostly based upon data from the Internet Amateur Astronomers Catalogue (IAAC) – at http://www.visualdeepsky.org/search.cgi.
Column 11 – The radio brightness at 0.2 m wavelength.
Column 12 – The redshift

Table 7.2 *Continued*

Name(s)	AGN type	RA_{2000}	Dec_{2000}	Constellation	Apparent integrated visual magnitude of AGN and galaxy	Apparent visual magnitude of AGN core	Angular size (seconds of arc)	Visibility class	Comments and observing notes (ON)	0.2 m radio brightness (Jy)	Redshift (z)
		h m s	° ′ ″								
4C+25.01	Q	00 19 40	+26 02 52	And	15.4	15.9		C		0.516	0.284
PG 0026+12	QSO	00 29 14	+13 16 04	Psc	14.78	15.41		B		0.007	0.142
NGC 193, 4C+03.01	RS	00 39.3	+03 20	Psc	12.3		100 × 100	C	ON – faint fuzzy disk seen at ×200 in a 0.3-m telescope	1.8	0.014
PG 0043+039	QSO	00 45 45	+04 11 04	Psc	15.88 (B)	16.0		C			0.384
PKS 0044+030	Q	00 47 06	+03 19 55	Psc	15.97 (B)	16.37		C		0.162	0.624

Table 7.2 *Continued*

Name(s)	AGN type	RA_{2000}	Dec_{2000}	Conste-llation	Apparent integrated visual magnitude of AGN and galaxy	Apparent visual magnitude of AGN core	Angular size (seconds of arc)	Visibil-ity class	Comments and observing notes (ON)	0.2 m radio bright-ness (Jy)	Red-shift (z)
NGC 253, C 65 PKS 0045–255	RDS	00 47.6	–25 17	Scl	7.1		1,500 × 440	C	Part of the Sculptor group of galaxies. Silver coin galaxy. The closest starburst galaxy (10 Mly, 3 Mpc). ON – Visible in binoculars, easily seen at ×50 in a 0.25-m telescope.	6.0	0.001
PG 0052+251	QSO	00 54 52	+25 25 39	Psc	15.42 (B)	15.43		C			0.155
PHL 909, PKS 0054+144	QSO	00 57 10	+14 46 11	Psc	15.71	16.32		C		0.002	0.171
NGC 383, 3C31, 4C+32.05	RDE	01 07.5	+32 24	Psc	11.8		140 × 110	C		5.4	0.016

NGC 404	LNR	01 09 27	+35 43 05	And	10.11	11.73	260 × 250	C	ON – Easily seen as a round and uniform shape at ×100 in a 0.12-m refractor.	0.003	0.001
S2 0109+22	HPQ	01 12 06	+22 44 39	Psc	15.5v	15.66		C		0.374	
NGC 547, 3C40, 4C–01.08	RE	01 26.0	–01 21	Cet	12.3		100 × 100	C	Twin with spiral galaxy NGC 545. Both galaxies the radio source 3C40. ON – Seen as a centrally-brightened round shape at ×250 in a 0.25-m telescope	5.2	0.018
NGC 612, PKS 0131–367	QSO	01 33 58	–36 29 36	Scl	13.2	13.2		A			0.030
3C48, 4C+32.08, NRAO 79	Q	01 37 41	+33 09 35	Tri	16.2v	16.2		C	The first quasar to be found, though not the first to be identified.	15.651	0.367

Table 7.2 *Continued*

Name(s)	AGN type	RA_{2000}	Dec_{2000}	Conste-llation	Apparent integrated visual magnitude of AGN and galaxy	Apparent visual magnitude of AGN core	Angular size (seconds of arc)	Visibil-ity class	Comments and observing notes (ON)	0.2 m radio bright-ness (Jy)	Red-shift (z)
NGC 660	LNR	01 43 02	+13 38 30	Psc	10.78	11.86	550 × 250	D	ON – Visible using averted vision in a 0.2-m telescope at ×200 or so.	0.473	0.003
NGC 676	Sy2	01 48 57	+05 54 25	Psc	10.99 (B)	10.5	260 × 90	C	ON – Easily seen as a centrally condensed, possibly slightly elongated, haze in a 0.5-m telescope.		0.005
Mrk 586	QSO	02 07 50	+02 42 56	Cet	15.41	15.41		C			0.155
3C66A	HPQ	02 22 40	+43 02 08	And	15.5v	15.21		C	RE 3C66B lies 6′ to the south	2.233	0.44?

3C66B, 4C+42.07	RE	02 23.3	+43 02	And	12.9			A	HPQ 3C66A lies 6′ to the north.	9.5	0.022
OD 160, AO 0235+163	HPQ	02 38 39	+16 37 00	Ari	15.50v	15.50		C		1.877	0.940
NGC 1044, 4C+08.11	RDS	02 41.1	+08 45	Cet	13.2		36 × 36	B		1.3	0.021
M 77, NGC 1068, 3C71	Sy2	02 42 41	–00 00 41	Cet	8.8	10.83	410 × 350	C	One of Seyfert's original galaxies. The brightest Sy 2 as seen from the Earth. Figs 3.4 and 7.1. ON – Small round haze in binoculars. Easily seen as an oval cloud with a bright core at ×100 in a 0.2-m telescope.	1.918	0.004
NGC 1058	Sy2	02 43 30	+37 20 27	Per	11.54	11.75	180 × 170	D		0.003	0.002
3C75, 4C+06.15	RDE	02 57.7	+06 03	Cet	13.1		54 × 24	B	Fig 4.6	6.0	0.024

Table 7.2 *Continued*

Name(s)	AGN type	RA_{2000}	Dec_{2000}	Constellation	Apparent integrated visual magnitude of AGN and galaxy	Apparent visual magnitude of AGN core	Angular size (seconds of arc)	Visibility class	Comments and observing notes (ON)	0.2 m radio brightness (Jy)	Redshift (z)
NGC 1218, 3C78, 4C+03.05	Sy1	03 08 26	+04 06 39	Cet	12.8	13.9	80 × 65	C		2.573	0.029
NGC 1265, 3C83.1, 4C+41.06	RE	03 18.2	+41 55	Per	12.5		130 × 110	D		8.4	0.025
NGC 1275, C 24, Per A, 3C84	Sy1.5	03 19 48	+41 30 42	Per	11.6	12.48	160 × 110	C	One of Seyfert's original six galaxies. ON – Easily seen as a bright, diffuse core with an extended halo at ×350 in a 0.35-m telescope	22.51	0.018
NGC 1316, For A	RDS	03 22.6	−37 14	For	8.8		430 × 310	C			0.006

3C88, 4C+02.10	LNR	03 27 54	+02 33 41	Tau	13.6	15.5	90 × 65	D	Also double radio source	4.739	0.030
NGC 1399, PKS 0336–355	RDE	03 38.8	–35.23	For	9.9		190 × 190	C	ON – uniform oval with a bright core. Easy in a 0.35-m telescope at ×200	2.5	0.005
PKS 0405–12	Q	04 07 48	–12 11 37	Eri	14.82v	14.86		B		2.368	0.574
3C110, 4C–05.17, NRAO 170	Q	04 17 17	–05 53 45	Eri	15.94v	15.94		C		0.691	0.773
BW Tau, 3C120, 4C+050.2	Sy1.5	04 33 11	+05 21 15	Tau	13.7	15.05	60 × 40	C	Named as a variable star	3.381	0.033
PKS 0521–36	Sy1	05 22 58	–36 27 31	Col	14.62v	14.62		B	This object is also classed as a blazar	15.56	0.055
3C147, NRAO 221	Q	05 42 36	+49 51 08	Aur	17.8v			D	Fig. 7.1.	22	0.545
PKS 0548–322	HPQ	05 50 42	–32 16 11	Col	15.5v	15.5		C		0.297	0.069
PKS 0637–75	QSO	06 35 47	–75 16 17	Men	15.75	15.75		C			0.651
NGC 2442	LNR	07 36 24	–69 31 48	Vol	11.22 (B)	11.2	360 × 330	E			0.005
PKS 0735+17	HPQ	07 38 07	+17 42 21	Gem	14.85v	16.22		B		1.102	0.424
PKS 0754+100	HPQ	07 57 07	+09 56 34	Cnc	14.5v	15.0		B		1.035	0.266

Table 7.2 *Continued*

Name(s)	AGN type	RA_{2000}	Dec_{2000}	Constellation	Apparent integrated visual magnitude of AGN and galaxy	Apparent visual magnitude of AGN core	Angular size (seconds of arc)	Visibility class	Comments and observing notes (ON)	0.2 m radio brightness (Jy)	Redshift (z)
B3 0754+394	QSO	07 58 00	+39 20 27	Lyn	14.36	14.36		B		0.012	0.096
PKS 0818–128	BLL	08 20 58	−12 58 59	Pup	15.5v	15.0		C		1.087	
3C206, NRAO 299, PKS0837–12	QSO	08 39 51	−12 14 34	Hya	15.76v	15.76		C			0.200
NGC 2639	LNR	08 43 38	+50 12 21	UMa	11.76	11.88	120 × 80	C		0.107	0.011
NGC 2663, PKS 0843–336	RE	08 45.1	−33 48	Pyx	12.3			A		2.2	0.008
SBS 0846+513	Q	08 49 58	+51 08 29	UMa	15.72v	18.8		C	This object is also classed as a blazar	0.351	0.584
NGC 2681	LNR	08 53 33	+51 18 48	UMa	10.3	11.33	230 × 210	C	ON – Centrally condensed bright elongated haze in a 0.3-m f10 telescope at ×300.	0.005	0.003

OJ 287	HPQ	08 54 49	+20 06 32	Cnc	14.0v	15.43		B		1.182	0.306
NGC 2655	LNR	08 55 39	+78 13 25	Cam	10.09	11.08	310 × 260	C	ON – Small, almost star-like but clearly seen in a 0.1-m f9 telescope at ×100	0.118	0.005
NGC 2768	Sy	09 11 38	+60 02 14	UMa	9.97	11.90	380 × 170	C	ON – Centrally condensed bright elongated haze in a 0.3-m f10 telescope at ×300.	0.012	0.005
Hya A, 3C 218	LNR	09 18 06	−12 05 43	Hya	13.6	14.8	50 × 50	C		23.38	0.054
NGC 2824, Mrk 394	Sy?	09 19 02	+26 16 11	Cnc	14.3 (B)	10.88	80 × 50	D		0.008	0.008
NGC 2787	LNR	09 19 19	+69 12 12	UMa	10.8	11.79	200 × 140	C	ON – Strongly centrally condensed, faint disk, easy in a 0.25-m telescope at ×200.	0.010	0.003

Table 7.2 *Continued*

Name(s)	AGN type	RA_{2000}	Dec_{2000}	Conste-llation	Apparent integrated visual magnitude of AGN and galaxy	Apparent visual magnitude of AGN core	Angular size (seconds of arc)	Visibil-ity class	Comments and observing notes (ON)	0.2 m radio bright-ness (Jy)	Red-shift (z)
NGC 2985	Sy1.9	09 50 22	+72 16 45	UMa	10.51	10.61	260 × 200	C	ON – Circular, centrally condensed faint haze in a 0.25-m f10 telescope at ×80.	0.019	0.004
M 81, NGC 3031	LNR	09 55 33	+69 03 55	UMa	6.93	11.63	1,540 × 850	C	Bode's galaxy ON – Large, bright, elongated and centrally condensed in a 0.4-m f4.5 telescope at ×150. Dark absorption lane may have been seen to the south-west.	0.624	0.000

M 82, NGC 3034, 3C231	STB	09 55.8	+69 41	UMa	8.4		670 × 280	C	The prototypical starburst galaxy. Figs 2.1 and 7.1. ON – Can be picked up in 0.1-m telescope. Elongated with patchy bright regions in a 0.3-m f10 telescope at ×200.	8.1	0.001
3C232, 4C+32.33, Ton 469	Q	09 58 21	+32 24 02	Leo	15.78v	15.78		C		1.229	0.533
Double quasar, Q0957+561A and B	Q	10 01 21	+55 53 56	UMa	17.0	16.98	6	D	The first gravitationally-lensed quasar to be found in 1979. The two equally bright components are separated by 6″. Fig. 7.1.	0.537	1.413
PG 1004+130, OL 107.7	Q	10 07 26	+12 48 57	Leo	15.15v	15.71		C		1.126	0.240

Table 7.2 *Continued*

Name(s)	AGN type	RA_{2000}	Dec_{2000}	Conste-llation	Apparent integrated visual magnitude of AGN and galaxy	Apparent visual magnitude of AGN core	Angular size (seconds of arc)	Visibil-ity class	Comments and observing notes (ON)	0.2 m radio bright-ness (Jy)	Red-shift (z)
Ton 34	QSO	10 19 55	+27 45 55	Leo	15.69	15.69		C			1.918
NGC 3227	Sy.5	10 23 31	+19 51 56	Leo	10.75	11.79	340 × 240	D	ON – Visible with averted vision in a 0.1-m telescope. Easier in a 0.25-m f6 telescope at ×50– circular and slight central condensation, forms a twin with NGC 3226.	0.083	0.003

NGC 3254	Sy2	10 29 20	+29 29 30	LMi	11.5	11.60	310 × 110	D	ON – Uniform, elongated haze with a slight central condensation, best seen using averted vision in a 0.25-m telescope at ×80.		0.005
NGC 3486	Sy2	11 00 24	+28 58 30	LMi	10.33	11.28	410 × 320	D	ON – Just glimpsed in a 0.1-m telescope. Round and uniform haze in a 0.3-m f10 telescope at ×100.	0.007	0.002
3C249.1, 4C+77.09, NRAO 363	QSO	11 04 14	+76 58 58	Dra	15.72v	15.72		C			0.311
Mrk 421	HPQ	11 04 27	+38 12 32	UMa	13.5v	12.9		A	ON – Visible in a 0.08-m telescope as a star-like object when at its brightest (12^m).	0.574	0.031

Table 7.2 *Continued*

Name(s)	AGN type	RA_{2000}	Dec_{2000}	Constellation	Apparent integrated visual magnitude of AGN and galaxy	Apparent visual magnitude of AGN core	Angular size (seconds of arc)	Visibility class	Comments and observing notes (ON)	0.2 m radio brightness (Jy)	Redshift (z)
4C+16.30, OM 109	Q	11 07 14	+16 27 52	Leo	15.7v	15.7		C		0.280	0.634
Triple QSO, PG 1115+080	QSO	11 18 17	+07 46 00	Leo	15.9	16.47	2 × 2	C	The second gravitationally lensed QSO to be found in 1980. Components B and C have magnitudes of 19 and 18.6. Component A was later found to be double, so this is actually a quadruple QSO.		1.722
PG 1116+215	QSO	11 19 09	+21 19 18	Leo	15.17 (B)	14.72		C		0.005	0.177

M 66, NGC 3627	LNR	11 20 15	+12 59 27	Leo	9.04	11.94	520 × 260	C	ON – Visible with averted vision in binoculars. Large and bright in a 0.2-m telescope. A 1-m telescope shows some structure – elongated central region and core plus bright patches and dark lanes.	0.434	0.002
NGC 3718	LNR	11 32 35	+53 04 04	UMa	10.53	10.61	520 × 270	D		0.011	0.003
Mrk 180	BLL	11 36 27	+70 09 24	Dra	14.49v	14.49		B		0.240	0.046
NGC 3941	Sy2	11 52 55	+36 59 11	UMa	11.36 (B)	11.62	230 × 150	D	ON – Easily seen in a 0.25-m f4.5 telescope at up to ×100 as a slightly elongated, centrally condensed object		0.003

Table 7.2 *Continued*

Name(s)	AGN type	RA_{2000}	Dec_{2000}	Constellation	Apparent integrated visual magnitude of AGN and galaxy	Apparent visual magnitude of AGN core	Angular size (seconds of arc)	Visibility class	Comments and observing notes (ON)	0.2 m radio brightness (Jy)	Redshift (z)
NGC 3982	Sy1.9	11 56 28	+55 07 30	UMa	11.74 (B)	11.70	150 × 130	C	ON – Visible as a faint circular haze in 0.3-m telescope at ×200.	0.004	0.003
4C+29.45, Ton 599	HPQ	11 59 32	+29 14 45	UMa	15.6v	14.41		C		1.953	0.729
NGC 4036	LNR	12 01 27	+61 53 45	UMa	10.58	11.20	270 × 120	C	ON – Can just be seen using averted vision with a 0.1-m telescope. Easily seen as an elongated centrally condensed nebula using a 0.35-m ×150.	0.011	0.005

GQ Com	QSO	12 04 42	+27 54 11	Com	15.51v	15.6		C	Named as a variable star		0.165
NGC 4151	Sy1.5	12 10 33	+39 24 21	CVn	10.38	11.85	350 × 260	D	One of Seyfert's original galaxies. ON – Easily seen as an elongated, centrally condensed nebulosity with additional structure in a 0.5-m telescope	0.331	0.003
NGC 4203	LNR	12 15 05	+33 11 50	Com	10.66	11.99	220 × 200	C	ON – Uniform circular haze with a star-like core at ×100 in a 0.25-m f4.5 telescope.	0.008	0.004
ON 325, B2 1215+30	HPQ	12 17 52	+30 07 01	Com	15.5v	15.62		C		0.406	0.130

Table 7.2 *Continued*

Name(s)	AGN type	RA_{2000}	Dec_{2000}	Constellation	Apparent integrated visual magnitude of AGN and galaxy	Apparent visual magnitude of AGN core	Angular size (seconds of arc)	Visibility class	Comments and observing notes (ON)	0.2 m radio brightness (Jy)	Redshift (z)
M 106, NGC 4258	Sy2	12 18 58	+47 18 14	CVn	8.31	11.65	1,090 × 470	C	One of the three galaxies (along with the Milky Way Galaxy and M 31) to possess a confirmed central black hole, whose mass is about 36 megasuns ON – Visible as an elongated nebulosity using binoculars and averted vision. Clearly seen as a strongly centrally condensed oval	0.792	0.002

									with additional structure in a 0.1-m telescope. A 0.6-m instrument reveals an extended halo about a third of a degree long and several brighter patches in the central region.		
M 59, NGC 4261, 3C270	LNR	12 19 23	+05 49 29	Vir	10.3	12.87	230 × 190	C	Fig. 5.5 ON – Oval, centrally condensed haze, easily seen using a 0.2-m f6 telescope at ×100	19.43	0.007

Table 7.2 *Continued*

Name(s)	AGN type	RA_{2000}	Dec_{2000}	Constellation	Apparent integrated visual magnitude of AGN and galaxy	Apparent visual magnitude of AGN core	Angular size (seconds of arc)	Visibility class	Comments and observing notes (ON)	0.2 m radio brightness (Jy)	Redshift (z)
NGC 4274	LNR	12 19 51	+29 36 50	Com	10.37	10.47	410 × 170	C	ON – Uniform oval nebulosity with star-like nucleus visible using a 0.25-m telescope at ×100.	0.009	0.003
NGC 4278	LNR	12 20 07	+29 16 51	Com	10.21	10.87	220 × 210	C	ON – Centrally condensed oval haze, easy using a 0.5-m telescope.	0.402	0.002
Mrk 205	Sy1	12 21 44	+75 18 37	Dra	14.5	15.24		B	Almost bright enough to be a QSO ON – Visible as a faint star in a 0.2-m telescope.		0.070

M 84, NGC 4374 3C272.1	Sy2	12 25 04	+12 53 13	Vir	9.3	12.31	560 × 60	B	ON – Easily seen in binoculars. Centrally condensed bright elliptical haze in 0.12-m and 0.2-m telescopes.	6.495	0.003
NGC 4395	Sy1.8	12 25 49	+33 32 48	CVn	10.15	10.27	770 × 660	E	The closest Seyfert galaxy (8 Mly, 2.5 Mpc). Fig. 3.6.	0.001	0.001
B2 1225+31	Q	12 28 25	+31 28 38	CVn	15.87	15.87		C	Note very high redshift	0.324	2.230
3C273, NRAO 400	Q	12 29 07	+02 03 07	Vir	12.86v	12.85		A	The brightest quasar as seen from the Earth and the first to be identified. Figs 3.14 and 4.3. ON – Possible, appearing as a star-like object, in 0.3-m and larger telescopes.	36.98	0.158

Table 7.2 *Continued*

Name(s)	AGN type	RA_{2000}	Dec_{2000}	Conste-llation	Apparent integrated visual magnitude of AGN and galaxy	Apparent visual magnitude of AGN core	Angular size (seconds of arc)	Visibil-ity class	Comments and observing notes (ON)	0.2 m radio bright-ness (Jy)	Red-shift (z)
M 87, NGC 4486 Vir A, 3C274	RDE	12 30 50	+12 23 28	Vir	8.6	12.86	520 × 60	A	Figs 3.2 and 7.1. ON – Perhaps visible in binoculars using averted vision. Easily seen using a 0.2-m telescope as a large round nebulosity brightening towards the center but otherwise featureless. Little more to be seen using larger instruments.	22.37	0.004

M 89, NGC 4552	Sy2	12 35 40	+12 33 23	Vir	9.81	11.20	590 × 60	B	ON – Easily seen in a 0.2-m telescope as a small, circular, centrally condensed haze.	0.113	0.001
M 58, NGC 4579	LNR	12 37 44	+11 49 05	Vis	9.78	11.72	320 × 260	C	ON – Just visible in binoculars. 0.2-m and larger telescopes show it as a uniform slightly elliptical nebulosity.	0.039	0.005

Table 7.2 *Continued*

Name(s)	AGN type	RA_{2000}	Dec_{2000}	Constellation	Apparent integrated visual magnitude of AGN and galaxy	Apparent visual magnitude of AGN core	Angular size (seconds of arc)	Visibility class	Comments and observing notes (ON)	0.2 m radio brightness (Jy)	Red-shift (z)
M 104, NGC 4594	Sy1.9	12 39 59	−11 37 23	Vir	8.3	9.25	530 × 250	B	Sombrero galaxy. Fig. 7.1. ON – Elliptical, centrally condensed object seen in a 0.2-m telescope. The central dark dust lane is difficult to detect visually and larger instruments will usually be needed.	0.085	0.002
NGC 4636	LNR	12 42 50	+02 41 16	Vir	9.56	11.84	370 × 300	C	ON – Large oval centrally condensed haze through a 0.55-m f4 telescope at ×200.	0.057	0.003

PG 1241+176	Q	12 44 11	+17 21 05	Com	15.38 (B)	15.9		C		0.362	1.273
NGC 4696, PKS 1246–410	LNR	12 48 49	–41 18 40	Cen	10.7	11.4	210 × 190	C		3.8	0.009
M 94, NGC 4736	LNR	12 50 53	+41 07 10	CVn	8.17	10.85	660 × 550	C	Has also been classed as a Sy1 and a starburst galaxy. ON – Just visible using averted vision in binoculars. Easily seen as a small, centrally condensed object using 0.1-m telescope. Larger instruments may show some faint structure.	0.276	0.001
NGC 4760, PKS 1250–102	RE	12 53.1	–10 29	Vir	12		110 × 110	C		0.9	0.014

Table 7.2 *Continued*

Name(s)	AGN type	RA_{2000}	Dec_{2000}	Constellation	Apparent integrated visual magnitude of AGN and galaxy	Apparent visual magnitude of AGN core	Angular size (seconds of arc)	Visibility class	Comments and observing notes (ON)	0.2 m radio brightness (Jy)	Redshift (z)
NGC 4782–3, 3C278	RDE	12 54 36	−12 33 48	Crv	11.7	11.6	100 × 100	C	Twin elliptical galaxies. Also classed as a LINER	8.2	0.014
PG 1254+047	QSO	12 57 00	+04 27 34	Vir	15.84 (B)	16.28		C			1.024
NGC 4945, C 83, PKS 1302–492	Sy	13 05 28	−49 28 03	Cen	9.2	14.4	1,200 × 260	D	ON – Seen as a bright elongated nebulosity in a 0.25-m f6 telescope.	6.6	0.002
PKS 1302–102 OP 106	Q	13 05 33	−10 33 20	Vir	15.23	15.23		C		0.700	0.286
PG 1307+085	QSO	13 09 47	+08 19 49	Vir	15.28 (B)	15.11		C			0.155
B2 1308+32	HPQ	13 10 29	+32 20 43	CVn	15.24	15.24		C		1.506	0.996
4C+07.32	RDE	13 16.3	+07 03	Vir	13.4		40 × 30	B		2.2	
Ton 153, CSO 873	QSO	13 19 56	+27 28 09	Com	15.98	15.98		C			1.014

NGC 5128, C 77 Cen A	RDE	13 25 28	−43 01 00	Cen	7.0	12.76	1,090 × 870	C	The closest double-lobed radio galaxy to the Earth (11 Mly, 3.5 Mpc) – Fig. 3.19 ON – Visible, including central dark lane, with difficulty in a 0.08-m telescope. Easily seen using a 0.25-m f6 telescope.	0.002
M 51, NGC 5194	LNR	13 29.9	+47 12	CVn	8.38		660 × 470	C	Whirlpool galaxy. Sometimes also classed as a weak Sy2 galaxy. Fig. 7.1. ON – Visible in binoculars. A 0.1-m telescope will show both galaxies and perhaps a hint of the spiral arms. The spiral shape is easily discerned in a 0.4-m telescope.	0.002

Table 7.2 *Continued*

Name(s)	AGN type	RA_{2000}	Dec_{2000}	Constellation	Apparent integrated visual magnitude of AGN and galaxy	Apparent visual magnitude of AGN core	Angular size (seconds of arc)	Visibility class	Comments and observing notes (ON)	0.2 m radio brightness (Jy)	Red-shift (z)
PG 1333+176	Q	13 36 02	+17 25 14	Com	15.64 (B)	16.23		C		0.033	0.554
IC 4296, PKS 1333–336	LNR	13 36 39	−33 57 58	Cen	10.6	12.99		A			0.012
M 83, NGC 5236	STB	13 37.0	−29 52	Hya	8.0		670 × 610	C	Pinwheel galaxy ON – Visible in binoculars. Easily seen in a 0.2-m telescope with m a star-like core.	2.6	0.001
PG 1351+64	Q	13 53 16	+63 45 45	Dra	14.84	14.28		B		0.027	0.088
NGC 5353	Sy?	13 53 27	+40 16 59	CVn	11.06	10.91	170 × 90	C	ON – A uniform, centrally condensed oval that can be seen using a 0.25-m telescope.	0.038	0.008

NGC 5371	Sy?	13 55 40	+40 27 43	CVn	10.75	11.93	260 × 220	D	ON – A uniform, oval haze that can be seen using a 0.25-m telescope.	0.009	0.007
PKS 1355–41	QSO	13 59 00	–41 52 53	Cen	15.81v	15.86		C			0.313
NGC 5419, PKS1400–337	RE	14 03.4	–33 58	Cen	12.4			A		0.8	0.014
Ton 182, PG 1402+261	QSO	14 05 16	+25 55 40	Boö	15.57 (B)	15.54		C		0.001	0.164
PG 1407+265	QSO	14 09 24	+26 18 21	Boö	15.73 (B)	15.74		C		0.009	0.944
NGC 4432, 3C296	RDS	14 16.9	+10 48	Boö	12.0		110 × 110	C		4.2	0.024
OQ 530	HPQ	14 19 47	+54 23 14	Boö	15.0v	15.65		C		0.584	0.152
Ton 202, B2 1425+26	Q	14 27 36	+26 32 14	Boö	15.68	15.68		C		0.345	0.366
S4 1435+63	Q	14 36 46	+63 36 38	Dra	15.0	16.66		C	Note very high redshift	0.878	2.060
PG 1444+407	QSO	14 46 46	+40 35 09	Boö	15.95 (B)	16.14		C			0.267
IC 1065, 3C305	Sy2	14 49 22	+63 16 14	Dra	13.4	15.71	80 × 65	D		2.942	0.041
4C+37.43, OR 321, B2 1512+37	Q	15 14 43	+36 50 51	Boö	15.5	17.04		C		0.872	0.371
3C317, 4C+07.40	Sy2	15 16 45	+07 01 17	Ser	13.5	14.17	130 × 65	D		4.536	0.035

Table 7.2 *Continued*

Name(s)	AGN type	RA_{2000}	Dec_{2000}	Constellation	Apparent integrated visual magnitude of AGN and galaxy	Apparent visual magnitude of AGN core	Angular size (seconds of arc)	Visibility class	Comments and observing notes (ON)	0.2 m radio brightness (Jy)	Redshift (z)
AP Lib	HPQ	15 17 42	−24 22 19	Lib	14.8v	14.8		B	Named as a variable star	1.983	0.049
NGC 5920, 3C318.1	RS	15 21.8	+07 40	Ser	13.6		85 × 65	D			0.046
PG 1522+101	QSO	15 24 25	+09 58 30	Ser	15.74 (B)	16.2		C			1.321
4C+14.60, OR 165	HPQ	15 40 50	+14 47 49	Ser	15.5v	17.3		C		1.483	0.605
Ton 256	Sy1	16 14 13	+26 04 16	CrB	15.41v	16.53		C	Almost bright enough to be a QSO	0.018	0.131
NGC 6166, 3C338	RE	16 28.6	+39 33	Her	12.0		140 × 110	C	ON – Just visible (averted vision may be needed) in a 0.3-m f10 telescope	3.2	0.030
PG 1630+377	QSO	16 32 01	+37 37 51	Her	15.96 (B)	16.33		C			1.471
PG 1634+706	QSO	16 34 29	+70 31 32	Dra	14.90 (B)	15.27		B		0.001	1.334
3C345, NRAO 513	HPQ	16 42 59	+39 48 37	Her	15.96v	16.62		C		6.599	0.595

Mrk 501	HPQ	16 53 52	+39 45 37	Her	13.88v	13.78		B	Fig. 3.13.	1.420	0.034
PG 1700+518	Q	17 01 25	+51 49 21	Dra	15.43 (B)	15.12		C		0.022	0.292
3C351	Q	17 04 41	+60 44 30	Dra	15.28v	15.43		C		3.176	0.371
I ZW 187	BLL	17 28 19	+50 13 10	Her	15.97v	15.97		C		0.220	0.055
KAZ 102	QSO	18 03 29	+67 38 09	Dra	15.78	15.78		C			0.136
IC 4729	Sy2	18 39 57	−67 25 32	Pav	13.3	11.80	100 × 80	D			0.014
3C402, 4C+50.49	RDE	19 41.7	+50 37	Cyg	12.9		80 × 80	C		5.0	0.025
Cyg A, 3C405	RDE	19 59.5	+40 44	Cyg	15.1			C	The second brightest radio source in the sky and the brightest extragalactic source. Fig. 3.18.	1,260	0.057
PKS 2040–267	RDE	20 43.7	−26 33	Cap	13.5			A		2.4	0.40
IC 5063, PKS 2048–572	Sy1	20 52 02	−57 04 07	Ind	12.0	13.6	110 × 90	C		2.1	0.011
PG 2112+059	QSO	21 14 53	+06 07 41	Equ	15.52 (B)	15.77		C		0.003	0.466
PKS 2128–12	Q	21 31 35	−12 07 04	Cap	15.98v	16.11		C		1.599	0.501
PKS 2135–14	Q	21 37 45	−14 32 56	Cap	15.53v	15.53		C		3.760	0.200
OX 169	Q	21 43 36	+17 43 49	Peg	15.73v	15.73		C		0.632	0.213
PKS 2153–69	Sy1	21 57 06	−69 41 23	Ind	13.8	13.79		B		30	0.027
PKS 2155–304	HPQ	21 58 52	−30 13 30	PsA	13.09v	13.09		A		0.424	0.116

Table 7.2 *Continued*

Name(s)	AGN type	RA_{2000}	Dec_{2000}	Conste-llation	Apparent integrated visual magnitude of AGN and galaxy	Apparent visual magnitude of AGN core	Angular size (seconds of arc)	Visibil-ity class	Comments and observing notes (ON)	0.2 m radio bright-ness (Jy)	Red-shift (z)
BL Lac	BLL	22 02 43	+42 16 40	Lac	14.72v	14.72		B	The archetype of the BL Lac objects. Named as a variable star	5.824	0.069
4C+31.63, B2 2201+31A	Q	22 03 15	+31 45 38	Peg	15.47v	15.58		C		2.497	0.297
NGC 7217	LNR	22 07 53	+31 21 33	Peg	10.2	11.92	220 × 190	C	ON – Easily seen as a bright round object like a globular cluster in a 0.4-m telescope	0.008	0.004
NGC 7236–7, 3C442	RDS	22 14.7	+13 50	Peg	13.5	60 × 60		A	Twin spiral galaxies	3.6	0.027
3C449, 4C+39.69	RDS	22 31.3	+39 19	Lac	13.3		85 × 85	D		3.8	0.018

NGC 7385 4C+11.71	RDE	22 49.8	+11 36	Peg	12.3		100 × 85	C	ON – A faint uniform round nebulosity that can be found with difficulty using a 0.25-m telescope.	2.4	0.024
4C+11.72, OY 186, PKS2251+ 11	Q	22 54 10	+11 36 38	Peg	15.82v	15.82		C		1.389	0.323
NGC 7410	LNR	22 55 01	−39 39 42	Gru	10.4	11.80	330 × 120	C		0.003	0.006
IC 1459	LNR	22 57 11	−36 27 44	Gru	10.01	11.85		A		1.203	0.005
NGC 7469	Sy1	23 03.3	+08 52	Peg	11.85		110 × 80	C	One of Seyfert's original galaxies. Fig. 3.4.		0.017
NGC 7503, 4C+07.61	RE	23 10.7	+07 35	Psc	13.8		55 × 55	C		1.7	0.044
NGC 7720, 3C465	Sy1	23 38 39	+27 01 52	Peg	12.6	13.3	110 × 90	C		7.460	0.029
4C+09.74, OZ 073.5, PKS 2344+ 09	Q	23 46 37	+09 30 45	Peg	15.97v	15.97		C		1.786	0.677

7.2 Imaging

Charge-couple device (CCD) cameras have revolutionized all aspects of astronomical imaging in the last two or three decades. Doubtless there are still many observers using photographic cameras on their telescopes and producing excellent results, but anyone who is serious about deep sky observing will want to use a CCD. The advantages of CCD cameras are their sensitivity – 100 times that of photographic emulsion – and their enormous dynamic range. The latter is a measure of the range of intensities that are accurately recorded by a detector before it saturates. For a photographic emulsion the dynamic range is around 100. Thus a photographic image of an AGN that shows its faint outer regions will have the center completely overexposed. Conversely a correct exposure for the AGN's core will not detect the fainter parts. A CCD by contrast has a dynamic range of up to one hundred thousand, so a single exposure will contain details of both the bright and faint parts of the object. An example of this is shown in Fig 7.1(v) where the central regions of a single CCD image of M 87 have been reproduced at reduced intensity so that its jet may be seen on the (low dynamic range) print along with the faint outer zone. The disadvantages of CCDs are that they are expensive and small, although both these problems are reducing as time goes by.

Any CCD camera from a webcam to an expensive digital camera can be attached to a telescope and will produce results of some sort. However imaging AGNs is likely to need exposures of several tens of seconds to a few minutes and a dedicated astronomical CCD camera is almost essential. These cameras are cooled to several tens of degrees below zero, which eliminates the overflow of electrons from one pixel to another, they can accommodate long exposures and often have a built-in guidance system to ensure that the images are sharp. Since CCDs' sensitivity peaks in the near infrared, a deep red filter, such as the Wratten #29, can reduce the effects of light pollution considerably. Astronomical CCD cameras will also often come with a software package

Figure 7.1 AGNs through a 0.4-m Schmidt–Cassegrain telescope, imaged using a STL1301E CCD camera. Log-stretching is used to display the images to their best advantage in print form.
(i) M 77 (Sy 2). A single 10-minute exposure through a Wratten #29 filter.
(ii) 3C147 (quasar). A single 5-minute exposure.

Figure 7.1 *Continued*
(iii) M 82 (starburst galaxy). Two 5-minute exposures through a Wratten #29 filter that have been combined together.
(iv) Double quasar (quasar). A single 5-minute exposure.

Figure 7.1 *Continued*
(v) M 87 (double-lobed radio galaxy). The central regions are displayed at reduced brightness in order to show the jet. Six 5-minute exposures that have been combined together.
(vi) M 104 (Sy 1.9).

Figure 7.1 *Continued*
(vii) M 51 (LINER).
(Images courtesy of Bob Forrest and the University of Hertfordshire Observatory.)

that enables noise to be reduced and several images to be added together plus other image processing options. There are a number of manufacturers of astronomical CCD cameras (see popular astronomy magazines for advertisements) each of whom has several different designs of instrument and whose designs are changing rapidly with time. We will therefore not go into further details of the cameras since they are certain to be out of date by the time that this book is published. However whichever CCD camera you choose or already own, then if it is used on a 0.2-m or larger telescope it should be possible to image all the objects listed in Table 7.1 that are visible from your observing site. As an encouragement to observers currently owning a CCD or contemplating purchasing one, Fig. 7.1 shows the results that an experienced observer can achieve with a CCD on a 0.4-m telescope.

Finally you do not have to own a CCD camera or even a telescope in order to observe AGNs – and with a 2-m telescope at that. The

Faulkes telescope project[44] provides access to a 2-m telescope based in Hawaii with a second instrument currently under construction in Australia. The telescope is equipped with professional-standard CCD cameras and is accessed remotely. UK and Irish-based astronomical societies can apply for time on the telescope free, whilst non-UK/Irish societies are asked to pay a fee (£50 per session at the time of writing). Further details, including how to apply to use the telescope, can be found at http://www.faulkes-telescope.com/.

[44]Telescopes in Education (TIE) is a somewhat similar US-based project, but currently only provides observing opportunities to schools.

Appendix 1 Bibliography

Web Sites

Given the efficacy of internet search engines and the rapidity of change on the internet, there is little point in listing most web sites that are currently of relevance to students of active galaxies since they are likely to have changed, disappeared or been bettered by the time that this book appears. The one exception to this rule that it is worth making, since it is likely to remain available and useful for many years, is the master list of identified AGNs maintained by the Centre de Données Stellaires (CDS) in Strasbourg:

A Catalogue of Quasars and Active Galactic Nuclei, 12th edition, M.-P. Véron-Cetty, P. Véron. The catalogue is available to any enquirer in electronic form only at the CDS. The web site address is http://cdsweb.u-strasbg.fr

At the time of writing the route is: select "Catalogues", followed by "Browse the list of catalogues", followed by "VII Non-stellar objects", followed by "(VII/248) Quasars and Active Galactic Nuclei".

Astronomical Magazines and Journals

Popular Level

Astronomy
Astronomy Now
Ciel et Espace
Journal of the British Astronomical Association
New Scientist
Practical Astronomy
Publications of the Astronomical Society of the Pacific
Scientific American
Sky and Telescope
Sky at Night

Research Level

Astronomical Journal
Astronomy and Astrophysics
Astrophysical Journal
Monthly Notices of the Royal Astronomical Association
Nature
Science

Almanacs, Star Atlases, Star and Other Catalogues

Astronomical Almanac, HMSO/US Government Printing Office (published annually).

Handbook of the British Astronomical Association, British Astronomical Association (published annually).

*Concise Catalogue of Deep Sky Objects:Astrophysical information for 500 galaxies, clusters and nebulae.*W.H. Finlay, Springer-Verlag, 2003
Norton's 2000.0, edited by I. Ridpath, Longman, 1998.
Sky Atlas 2000.0, W. Tirion, Sky Publishing Corporation, 2000.
Sky Catalogue 2000, Volumes 1 and 2, A. Hirshfield and R.W. Sinnott, Cambridge University Press, 1992.
Yearbook of Astronomy, Macmillan, published annually.

Reference Books

Allen's Astrophysical Quantities, 4th edition, Ed. A. N. Cox, AIP press/Springer, 1999.
Encyclopedia of Astronomy and Astrophysics, edited by P Murdin, *Nature* and IoP Publishing, 2001.

Practical Astronomy Books

Analysis of Starlight; 150 Years of Astronomical Spectroscopy, J.B. Hearnshaw, Cambridge University Press, 1987.
Art and Science of CCD Astronomy, D. Ratledge, Springer-Verlag, 1997.
Astronomical Equipment for Amateurs, M. Mobberley, Springer-Verlag, 1999.
Astronomical Spectroscopy, C.R. Kitchin, Adam Hilger, 1995.
Astronomy with Small Telescopes, S.F. Tonkin, Springer-Verlag, 2001.
Astrophysical Techniques, C.R. Kitchin, 4th edition, IoP press, 2003,
Cambridge Deep-Sky Album, J. Newton and P. Teece, Cambridge University Press, 1983
Deep Sky Companions: The Messier objects, D.H. Levy, Cambridge University Press,, 1998.
Deep Sky Observer's Year, G. Privett and P. Parsons, Springer-Verlag, 2001.
Deep Sky Observing, S.R. Coe, Springer-Verlag, 2000.
Field Guide to the Deep Sky Objects, M. Inglis, Springer-Verlag, 2001.
Illustrated Dictionary of Practical Astronomy, C.R. Kitchin, Springer-Verlag, 2002.
Messier's Nebulae and Star Clusters, K.G. Jones, Cambridge University Press, 1991.

Observer's Sky Atlas, E. Karkoschka, Springer-Verlag, 1999.
Observer's Year, P. Moore, Springer-Verlag, 1998.
Observing Handbook and Catalogue of Deep Sky Objects, C. Luginbuhl and B. Skiff, Cambridge University Press, 1990.
Observing the Caldwell Objects, D. Ratledge, Springer-Verlag, 2000.
Photo-Guide to the Constellations, C.R. Kitchin, Springer-Verlag, 1997.
Seeing Stars, C. Kitchin and R. Forrest, Springer-Verlag, 1998
Star Gazing through Binoculars: A Complete Guide to Binocular Astronomy, S. Mensing, TAB, 1986.
Star Hopping: Your Visa to the Universe, R.A. Garfinkle, Cambridge University Press, 1993.
Using the Meade ETX, M. Weasner, Springer-Verlag, 2002.

Introductory Books

Astronomy: Principles and Practice, A.B. Roy and D. Clark, Adam Hilger, 1988.
Astronomy through Space and Time, S. Engelbrektson, W.C. Brown, 1994.
Eyes on the Universe, P. Moore, Springer-Verlag, 1997.
Introductory Astronomy, K. Halliday, John Wiley, 1999.
Introductory Astronomy and Astrophysics, M. Zeilik, S.A. Gregory and E.vP. Smith, Saunders, 1992.
Unfolding our Universe, I. Nicolson, Cambridge University Press, 1999.
Universe, R.A. Freedman and W.J. Kaufmann III, WH Freeman, 2001.

Galaxies and Active Galaxies

Active Galactic Nuclei, R.D. Blandford, H. Netzer, L. Woltjer, Springer-Verlag, 1990
Active Galactic Nuclei, Ed: J.E. Dyson, Manchester University Press, 1985
Active Galactic Nuclei, I. Robson, John Wiley, 1996
Active Galactic Nuclei: From the Central Black Hole to the Galactic Environment, J.H. Krolik, Princeton University Press, 1999
Active Galactic Nuclei in their Cosmic Environment, Eds: B. Rocca-Volmerange, H. Sol, EDP Sciences, 2001
Active Galaxies, Open University Worldwide, 2002.

Bibliography

Advanced Lectures on the Starburst–AGN Connection: Tonantzintla, Puebla, Mexico, 26–30 June 2000, Eds: I. Aretxaga, R. Mujica, D. Kunth, World Scientific Publishing, 2001

An Introduction to Active Galactic Nuclei, B.M. Paterson, Cambridge University Press, 1997

An Introduction to Galaxies and Cosmology, M.H. Jones, R.J. Lamborne, Cambridge University Press, 2004

Astrophysics of Gas Nebulae and Active Galactic Nuclei, G.J. Ferland, D.E. Osterbrock, University Science Books, 2005

Colliding Galaxies: The Universe in Turmoil, B. Parker, Kluwer Academic Press, 1990

Galactic Astronomy, J. Binney, M. Merrifield, Princeton University Press, 1998

Galaxies in the Universe: An introduction, L.S. Sparke, J.S. Gallagher, Cambridge University Press, 2000

Gravity's Fatal Attraction: Black Holes in the Universe, M. Begelman, M. Rees, Scientific American Library, 1996.

Nearly Normal Galaxies, Ed: S.M. Faber, Springer-Verlag, 1987

Quasars and Active Galactic Nuclei: An introduction, A.K. Kembhavi, J.V. Narlikar, Cambridge University press, 1999

Quasars, Redshifts and Controversies, H.C. Arp, Cambridge University Press, 1989

Seeing Red: Redshifts, Cosmology and Academic Science, H.C. Arp, Aperion press, 1998

Starburst Galaxies: Near and Far, Eds: L. Tacconi, D. Lutz, Springer-Verlag, 2001

Physics of Active Galactic Nuclei, Eds: W.J. Duschl, S.J. Wagner, Springer-Verlag, 1992

Radio Galaxies, A.G. Pacholczyk, Pergamon Press, 1977

The Edge of Infinity: Supermassive Black Holes in the Universe, F. Melia, Cambridge University Press, 2003

The Milky Way Galaxy, L.S. Marochnik, A.A. Suchkov, Gordon and Breach, 1995

The Nearest Active Galaxies, Ed: J. Beckman, Kluwer Academic Press, 1993

Theory of Black Hole Accretion Disks, Eds: M.A. Abramowicz, G. Bjvrnsson, J. Pringle, Cambridge University Press, 1999

The Structure and Evolution of Galaxies, S. Phillipps, John Wiley, 2005

The World of Galaxies, Eds: H.C. Corwin, L. Bottinelli, Springer-Verlag, 1989

Windows on Galaxies, Eds: G Fabbiano, J.S. Gallagher, A. Renzini, Kluwer Academic Press, 1990

Relativity and Quantum Mechanics

A First Course in General Relativity, B.F. Schutz, Cambridge University Press, 1985

General Relativity: An Introduction for Physicists, M. Hobson, G. Efstathiou, A. Lasenby, Cambridge University Press, 2006

Gravity from the Ground Up: An Introductory Guide to Gravity and General Relativity, B. Schutz, Cambridge University Press, 2003

Introductory Quantum Mechanics, R.L. Liboff, Benjamin Cummings, 2002

Quantum Mechanics: A Modern and Concise Introductory Course, D.R Bes, Springer-Verlag, 2004

Appendix 2
Glossary of Types of Galaxies and AGNs

For the newcomer, the numerous classes of AGNs and the other types of galaxy, together with their often confusing names can be bewildering. Here a brief description or definition is given as a handy source of quick reference.

Active Galaxy

A galaxy that has higher levels of activity within it than those occurring inside the average galaxy. The activity may take the form of high rates of star formation (starburst galaxies, etc.) or of non-thermal emissions from a small central core of the galaxy (AGNs, etc.).

AGN (Active Galactic Nucleus)

A general term for a galaxy where a significant amount, perhaps most, of the emitted energy comes from non-stellar sources within and around its central region. The optical and ultraviolet spectra of most AGNs contain very broad, strong emission lines and they may also have strong radio, infrared and/or x-ray emissions. There are numerous sub-types of AGN whose details are also listed within this glossary.

BALQSO (Broad-Absorption-Line Quasi-Stellar Object)

A QSO whose spectrum contains broad intrinsic absorption lines as well as emission lines.

Barred-Spiral Galaxy

See spiral galaxy.

BL Lac Object

One of a rare group of AGNs named after what was initially thought to be a variable star in the constellation of Lacerta. BL Lac objects have a featureless but strongly polarized spectrum. They vary in brightness by large amounts on timescales sometimes of just a few hours. They also emit strongly at radio wavelengths. Along with OVV quasars they form the Blazar type of AGN.

Blazar

A generic name for BL Lac objects and HPQs (including the OVV quasars). Although these are different types of AGN, their observational properties are sufficiently similar to make the class a useful concept.

BLR (Broad-Line Region)

A part of an AGN that produces emission lines with Doppler-broadening widths of several thousands of kilometers per second such as may be found in BLRGs and type-1 Seyfert galaxies.

BLRG (Broad-Line Radio Galaxy)

A galaxy whose spectrum is very similar to that of a Seyfert 1 but which emits strongly at radio wavelengths.

Buried Quasar/QSO

See type-2 quasars/QSOs.

Core-Dominated Radio Galaxy

A radio galaxy in which the radio emission comes mainly from a compact central source. The radio spectrum is nearly flat, and many of these galaxies are BL Lacs or OVV quasars.

Double-Lobed Radio Galaxy

A radio galaxy in which the radio emission comes mainly from two very large extended regions, called lobes, on either side of the galaxy. The lobes can be up to 10 Mly (3 Mpc) across, although sizes around

100,000 ly (30 kpc) are more common. Most double-lobed radio galaxies are giant elliptical galaxies.

Early-Type Galaxy

An elliptical galaxy. The name derives from the manner in which Hubble's classification system for galaxies places the elliptical galaxies on the left (or early in) the sequence of galactic types. It has no significance with regard to the age or evolutionary position of a galaxy.

Elliptical Galaxy

A galaxy that is a more or less smooth, symmetric, collection of stars. Their shapes can range from spherical to an ellipsoid with a major axis three times larger than its minor axis. They contain little gas or dust and in most cases the stars are relatively old, small and cool. They range in size from a few thousand ly (around 1 kpc) to over 100,000 ly (30 kpc) and in mass from 10 million to 10 trillion solar masses. Current ideas suggest that they are formed during the collision and merger of two large spiral galaxies.

Fanaroff–Riley Class

A classification system for radio galaxies. The galaxies are divided into the FR1 and FR 2 types with FR 1s having lower radio brightnesses and radio lobes that darken towards their edges. FR 2 galaxies are more luminous and their radio lobes brighten towards their edges or ends.

FR Class 1

See Fanaroff–Riley class.

FR Class 2

See Fanaroff–Riley class.

FIRG (Far InfraRed Galaxy)

A galaxy with strong emission in the far infrared. Many of these galaxies turn out to be Seyfert 2s, others are probably not true AGNs. The source of their infrared emission is not clear although it may be dust heated by hot stars or even a buried quasar. The brightest and most extreme examples of FIRGs are the ULIRGs.

Flocculent Spiral Galaxy

See spiral galaxy.

Grand Design Spiral Galaxy

See spiral galaxy.

HBL (High-Frequency BL Lac Object)

A synonym for x-ray loud BL Lacs (XBLs).

Hidden Quasar/QSO

See type-2 quasars/QSOs.

H II Region Galaxy

Another name for a starburst galaxy.

Host Galaxy

A relatively normal galaxy within which there is an AGN. Both spiral and elliptical galaxies can be hosts, and sometimes there is evidence for rapid star formation within the galaxy. High-luminosity AGNs are associated with high-luminosity hosts, but lower luminosity AGNs can be found within galaxies of any brightness. Host galaxies are often very difficult to observe, being swamped by the much brighter AGN.

HPQ (High-Polarization Quasar)

A quasar with a polarization in optical region of 3% or more. The OVV quasars may sometimes be regarded as a subset of the HPQs though there is little real difference between them. HPQs and BL Lacs are combined to form the Blazar class of AGNs.

IRAS Galaxy

A galaxy discovered by the Infrared Astronomy Satellite (IRAS) to have very strong emission in the far infrared (60–100 μm). They are also called FIRGs and ULIRGs.

Irregular Galaxy

A galaxy with a poorly defined, unsymmetrical or amorphous shape and structure. The group sub-divides into the Irr I and Irr II groups. Irr I galaxies are generally fairly small, being between about 3,000 ly (1,000 pc) to 30,000 ly (10,000 pc) across and with masses usually less than 3,000 million solar masses. They contain much gas and dust and usually show signs of rapid star formation such as groups of hot young stars (OB associations) and H II regions. Irr II galaxies are basically those that cannot be classified into any other group. They tend to be amorphous or to have a disturbed appearance and to show signs of powerful internal activity or of a recent interaction or a merger with another galaxy.

Irr I Galaxy

See Irregular galaxy.

Irr II Galaxy

See Irregular galaxy.

Lacertid

A synonym for a BL Lac object.

Late-Type Galaxy

A spiral galaxy. The name derives from the manner in which Hubble's classification system for galaxies places the spiral galaxies on the right (or late in) the sequence of galactic types. It has no significance with regard to the age or evolutionary position of a galaxy.

LBL (Low-Frequency BL Lac Object)

A synonym for radio-loud BL Lacs (RBLs).

Lenticular Galaxy

A galaxy with a nucleus surrounded by a flattened disk, but with no sign of spiral arms.

LINER Galaxy (Low-Ionization Nuclear Emission Region)

A galaxy, usually spiral, whose nucleus has emission lines present in its spectrum. Unlike a true AGN though, the emission lines in a LINER's spectrum arise from elements in low states of ionization. The emission lines probably result from interstellar matter heated by shock waves or very hot stars. Evidence of LINERs may be found in up to a third of spiral galaxies and while some may not be true AGNs, others may be low-intensity Seyfert galaxies. The term "Seyfert 3" is used by some workers as a synonym for LINER.

LINER 1

A LINER galaxy whose spectrum has weak broad components to its emission lines.

LINER 2

A LINER galaxy whose spectrum has only narrow emission lines, with no sign of any broader components.

LLAGN (Low-Luminosity Active Galactic Nucleus)

An AGN whose absolute visual magnitude is fainter than −23. The class includes Seyfert galaxies, BLRGs, NLRGs, LINERS, ULIRGs and starburst galaxies.

Lobe-Dominated Radio Galaxy

A synonym for a double-lobed radio galaxy.

LPQ (Low-Polarization Quasar)

A quasar with a polarization in optical region of less than 3% (i.e. a normal quasar).

Markarian Galaxy

A galaxy discovered during a survey conducted by Benjamin Markarian between 1962 and 1981. The survey used the 1.3-m Schmidt camera of the Byurakan observatory in Armenia together with an objective prism to find galaxies with strong emission in the blue region of the spectrum. Of the 1,500 galaxies discovered, about 10% were Seyfert galaxies and the remainder mostly starburst galaxies.

NLR (Narrow-Line Region)

A part of an AGN that produces emission lines with Doppler-broadening widths of a few hundred to a thousand kilometers per second such as may be found in NLRGs and type-2 Seyfert galaxies.

NLRG (Narrow-Line Radio Galaxy)

A galaxy whose spectrum is very similar to that of a Seyfert 2 but which emits strongly at radio wavelengths.

NLS1 (Narrow-Line Seyfert 1 galaxy)

See Seyfert Galaxy.

NLXG (Narrow-Line X-ray Galaxy)

A galaxy that is often similar to a Seyfert 2 but which emits strongly at x-ray wavelengths.

N-Type Galaxy

A type of galaxy identified early on in the development of research into AGNs. They are now recognized to be distant Seyfert galaxies, and the class is no longer much used.

OVV Quasar (Optically Violently Variable Quasar)

One of a small group of quasars that emit strongly at radio wavelengths and have flat, polarized, spectra. They resemble BL Lac objects except that their spectra contain wide and intense emission lines. Their variability is irregular with periods of large changes in luminosity over timescales of days or weeks, interspersed with long periods of quiescence. The group is best regarded as a subset of the HPQs. Along with BL Lac objects they form the Blazar type of AGN.

Peculiar Galaxy

A galaxy that is recognizably one of Hubble's classes of galaxy, but with differences or additional features that place it apart from classical examples of that type of galaxy.

Polar-Ring Galaxy

A rare type of galaxy that is surrounded by a ring of stars and other material. Polar ring galaxies differ from ring galaxies in that the ring is aligned at right angles to the disk of the galaxy (i.e. through the poles of the galaxy's rotation). The central galaxies are usually lenticular. Polar ring galaxies are thought to arise through glancing or asymmetric galactic collisions and mergers.

QSO (Quasi-Stellar Object)

A very distant compact object whose optical spectrum contains strong emission lines superimposed upon a non-thermal (i.e. non-stellar) continuum. QSOs are typically 100 times brighter than the most luminous classical galaxies with an absolute visual magnitude brighter than −23. Their radio luminosities are comparable with or only slightly stronger than those of classical galaxies. They can be found at distances ranging from 800 Mly (250 Mpc) to 13,000 Mly (4,000 Mpc) although most are around 12,000 Mly (3,500 Mpc) away from us. Their luminosities can vary on timescale as short 10 days or so, though most vary in brightness over a few months or years. QSOs are the radio-quiet equivalent of quasars. Commonly "quasar" and "QSO" are treated as synonyms, but in this book the terms are kept for radio-loud and radio-quiet objects respectively. QSOs outnumber quasars by about ten to one. See also type-1 and type-2 quasars/QSOs.

Quasar

A radio source whose optical counterpart is extremely compact and luminous. The name derives from "quasi-stellar radio source". The optical spectrum contains strong emission lines superimposed upon a non-thermal (i.e. non-stellar) continuum. Quasars are typically 100 times brighter than the most luminous classical galaxies with an absolute visual magnitude brighter than –23. They can be found at distances ranging from 800 Mly (250 Mpc) to 13,000 Mly (4,000 Mpc) although most are around 12,000 Mly (3,500 Mpc) away from us. Their luminosities can vary on timescale as short 10 days or so, though most vary in brightness over a few months or years. Quasars are the radio-loud equivalent of QSOs. Commonly "quasar" and "QSO" are treated as synonyms, but in this book the terms are kept for radio-loud and radio-quiet objects respectively. The number of quasars is about 10% that of QSOs. See also type-1 and type-2 quasars/QSOs.

Quasi-Stellar Object

See QSO.

Quasi-Stellar Radio Source

See quasar.

Radio Galaxy

A galaxy that emits much more energy in the radio region than does a classical galaxy. Classical galaxies have radio luminosities of around 10^{32} W; a radio galaxy can be up to a million times brighter. Radio galaxies sub-divide into core-dominated objects like BL Lacs and OVV quasars and double-lobed objects that are mostly elliptical galaxies and quasars.

Radio-Loud BL Lac Object (RBL)

A BL Lac object with strong emission over the radio region and only weak emission at x-ray wavelengths. This was the first type of BL Lac object to be found and includes BL Lac itself.

Ring Galaxy

A galaxy somewhat resembling a spiral galaxy, but within which the stars in the disk are collected into a circular or elliptical ring. The ring is usually separate from and surrounds the nucleus, but in some examples the nucleus is missing or seems to be incorporated into the ring. Numerical modeling suggests that ring galaxies are the result of a "bullseye" collision between a small, compact galaxy and a spiral galaxy (i.e. a collision in which the smaller galaxy comes in on a trajectory that is perpendicular to the plane of the spiral galaxy and in which it hits the spiral galaxy at or close to the center of its nucleus).

RLAGN (Radio-Loud Active Galactic Nucleus)

An AGN that emits strongly at radio wavelengths, such as quasars and BL Lac objects.

RLQ (Radio-Loud Quasar)

Synonym for a quasar (as defined in this book).

RQAGN (Radio-Quiet Active Galactic Nucleus)

An AGN that emits at a similar level to a classical galaxy, or only slightly more strongly, at radio wavelengths, such as QSOs and Seyfert galaxies.

RQQ (Radio-Quiet Quasar)

Synonym for a QSO (as defined in this book).

SCUBA Galaxy

See sub-millimeter galaxy.

Seyfert Galaxy

A member of the group of galaxies first recognized by Carl Seyfert in 1943 whose nuclei are small and very luminous with spectra containing broad, strong emission lines. Most Seyfert galaxies are spirals and they sub-divide into classes 1 and 2 (Sy 1, Sy 2) on the basis of the widths of the emission lines. The spectra of the nuclei of Seyfert 1 galaxies have broad emission lines due to hydrogen, but much narrower forbidden lines from heavier elements. For Seyfert 2 nuclei the widths of all the emission lines are similar and their widths lie between those of the broad and narrow Seyfert 1 lines. Further sub-classes of Seyferts (Sy 1.5, Sy 1.8, etc) are distinguished on the basis of the relative strengths of the broad and narrow components of the Hβ line. Seyferts are probably fainter versions of the QSOs and some of the low-intensity Seyferts may be classed as LINERs. In fact some workers use the name Seyfert 3 in place of LINER.

Narrow-Line Seyfert 1 galaxies (NLS1s) have recently been recognized as a sub-class of the Seyfert 1s. Their broad emission lines are less than 2,000 km/s in width (compared with up to 10,000 km/s for standard Seyfert 1s) but the ratio of the strengths of the [O III] 500.7 and H I 486.1 spectrum lines is lower than in Seyfert 2 galaxies and there are other differences that distinguish them from Seyfert 2s.

Seyfert Galaxy Type 1

See Seyfert galaxy.

Seyfert Galaxy Type 2

See Seyfert galaxy.

Seyfert Galaxy Type 3

A synonym for LINER – see Seyfert galaxy and LINER.

Sy 1 Galaxy

See Seyfert galaxy.

Sy 2 Galaxy

See Seyfert galaxy.

Sy 3 Galaxy

A synonym for LINER – see Seyfert galaxy and LINER.

Shell Galaxy

An elliptical galaxy that is surrounded by one or more shells of stars. The shells are usually incomplete and are thought to be the result of a recent collision with another, smaller, galaxy.

Spiral Galaxy

A galaxy whose structure includes a clear spiral pattern. The galaxies usually have a near-spherical central nucleus, surrounded by a flattened disk containing the spiral structure. Grand design spiral galaxies have two, sometimes three or four, well-defined spiral arms that merge with the nucleus at their inner ends. Flocculent spiral galaxies contain many short spiral segments that give the galaxy an overall spiral appearance, but individual arms within the structure are often not clearly identifiable. Barred spiral galaxies have a linear structure, or bar, running across the nucleus with the spiral arms emerging from the ends of the bar. Improved observations show that many, perhaps all, spiral galaxies contain such bars, but in a lot of cases they are small or hidden within the nucleus and so difficult to observe. Spiral galaxies tend to be large – up to 300,000 ly (100 kpc) across and with masses up to a trillion times that of the Sun. The disks of spiral galaxies contain many massive, hot young stars and thus tend to have a bluish tinge on color images. Their nuclei, by contrast, are uncolored, yellowish or even with a reddish tinge, since the stars therein are older, smaller and cooler.

Starburst Galaxy

A galaxy in which highly intense star formation is occurring whose rate may be up to a thousand times that of a classical galaxy. The star-forming region may be within the nucleus and the energy produced by the stars absorbed and re-emitted by dust. The star-forming region is then not seen directly but its presence inferred from the galaxy's strong infrared emission (see FIRGs, ULIRGs and IRAS galaxies). In other starburst galaxies, star-forming regions are found throughout the galaxy and can be seen directly. These galaxies often show signs of a recent collision or merger with another galaxy that has initiated the upsurge in the rate of star formation. Although sometimes classified as active galaxies, starburst galaxies are not true AGNs since their energy emission arises largely from stars.

Sub-Millimeter Galaxy

An extremely distant starburst galaxy, initially detected at sub-millimeter wavelengths by the SCUBA (Sub-millimeter Common User Bolometer Array) instrument on the James Clerk Maxwell telescope. The galaxies are bright in the infrared and so may also be classed as FIRGs.

Super-Starburst Galaxy

A synonym for a ULIRG.

Type-1 Quasar/QSO

The normal type of quasar or QSO with optical spectra similar to those of the type-1 Seyfert galaxies. Type-1 QSOs are thought to be the high-luminosity equivalents of type-1 Seyferts and type-1 quasars the equivalents of BLRGs.

Type-2 Quasar/QSO

Hypothesized high-luminosity analogues of the NLRGs and type-2 Seyfert galaxies. Examples of these objects have yet to be discovered with certainty, however a few relatively normal looking galaxies that have strong hard x-ray emission from their nuclei have been suggested as possible examples of type-2 QSOs. It has also been suggested that ULIRGs could be type-2 QSOs. They are also sometimes called hidden or buried quasars or QSOs.

ULIRG (Ultraluminous Infrared Galaxy)

The most luminous examples of FIRGs with far infrared brightnesses exceeding a trillion solar luminosities. They show evidence of being two colliding or merging galaxies and it is suggested that the infrared emission comes from dust heated by radiation from the resulting starbursts. However it is also possible that they contain a buried quasar or QSO that heats the dust. They have been proposed as examples of the elusive type-2 quasars or QSOs.

X-ray Galaxy

A galaxy that has much stronger x-ray emission from its nucleus than that which comes from classical galaxies.

X-ray Loud BL Lac Object (XBL)

A BL Lac object with strong emission over the x-ray region and weaker than normal (for BL Lac objects) emission at radio wavelengths.

Appendix 3
Acronyms

Acronym	Meaning
2dF	Two-Degree Field (of the AAT)
2MASS	Two Micron All Sky Survey
3C	Third Cambridge catalogue of radio sources
6dF	Six Degree Field (of the UK Schmidt camera)
AAO	Anglo-Australian Observatory
AAT	Anglo-Australian Telescope
AGN	Active Galactic Nucleus
ASCA	Advanced Satellite for Cosmology and Astrophysics
BALQSO	Broad-Absorption-Line Quasi-Stellar Object
BBB	Big Blue Bump

BLR	Broad-Line Region
BLRG	Broad-Line Radio Galaxy
CCD	Charge-Coupled Device (or Detector)
DRAGN	Double Radio source Associated with a Galactic Nucleus
e-m	electromagnetic
ENLR	Extended Narrow-Line Region
ESA	European Space Agency
ESO	European Southern Observatory
EUV	Extreme Ultraviolet
FIR	Far InfraRed
FIRG	Far InfraRed Galaxy
FR	Fanaroff–Riley class
FSC	Faint Source Catalogue (IRAS)
GMC	Giant Molecular Cloud
GRS	Galaxy Redshift Survey (by the AAT's 2dF)
HBL	High-frequency BL Lac object
HPQ	High-Polarization Quasar
HST	Hubble Space Telescope
IRAS	InfraRed Astronomy Satellite
ISO	Infrared Space Observatory
laser	Light Amplification by Stimulated Emission of Radiation
LBL	Low-frequency BL Lac object
LINER	Low-Ionization Nuclear Emission Region
LLAGN	Low-Luminosity Active Galactic Nucleus
LMC	Large Magellanic Cloud
LPQ	Low-Polarization Quasar
LPR	Light Pollution Rejection filter
maser	Microwave Amplification by Stimulated Emission of Radiation
MERLIN	Multi-Element Radio-Linked Interferometer Network
MIR	Mid InfraRed
MKN	Markarian
Mrk	Markarian
MWBR	MicroWave Background Radiation

NGC	New General Catalogue of nebulae and clusters of stars (J.L.E. Dreyer, 1888)
NIR	Near InfraRed
NLR	Narrow-Line Region
NLRG	Narrow-Line Radio Galaxy
NRAO	National Radio Astronomy Observatory
NLS1	Narrow-Line Seyfert 1 galaxy
OVV	Optically Violently Variable quasar
PKS	ParKeS radio sources catalogue
QSO	Quasi-Stellar Object
Quasar	Derived from quasi-stellar radio source
RBL	Radio-Loud BL Lac objects
RLAGN	Radio-Loud Active Galactic Nucleus
RLQ	Radio-Loud Quasar (= quasar)
RQAGN	Radio-Quiet Active Galactic Nucleus
RQQ	Radio-Quiet Quasar (= QSO)
SCUBA	Sub-millimeter Common User Bolometer Array
SDSS	Sloan Digital Sky Survey
SED	Spectral Energy Distribution
SMC	Small Magellanic Cloud
TIE	Telescopes In Education
ULIRG	Ultraluminous InfraRed Galaxy
VLA	Very Large Array
VLBI	Very Long Baseline Interferometry
VLT	Very Large Telescope (at ESO)
WIMP	Weakly Interacting Massive Particle
WMAP	Wilkinson Microwave Anisotropy Probe
XBL	X-ray Loud BL Lac objects

Appendix 4
SI and Other Units

SI Prefixes

Prefix	Multiplier	Symbol
atto	10^{-18}	a
femto	10^{-15}	f
pico	10^{-12}	p
nano	10^{-9}	n
micro	10^{-6}	μ
milli	10^{-3}	m
centi	10^{-2}	c (not recommended)
deci	10^{-1}	d (not recommended)
deca	10^{1}	da (not recommended)

Prefix	Multiplier	Symbol
hecto	10^2	h (not recommended)
kilo	10^3	k
mega	10^6	M
giga	10^9	G
tera	10^{12}	T
peta	10^{15}	P
exa	10^{18}	E

SI Units

Physical quantity	Unit	Symbol
angle	radian	rad (= $1/(2\pi)$ of a circle ≈ 57.2958°)
capacitance	farad	F ($s^4 A^2 m^{-2} kg^{-1}$)
electric charge	coulomb	C (A s)
electric current	ampere	A
electrical resistance	ohm	Ω ($m^2 kg\, s^{-3} A^{-2}$)
energy	joule	J ($m^2 kg\, s^{-2}$)
force	newton	N ($kg\, m\, s^{-1}$)
frequency	hertz	Hz (s^{-1})
length	meter	m
luminous intensity	candela	cd
magnetic flux density	tesla	T ($kg\, s^{-2} A^{-1}$)
mass	kilogram	kg
power	watt	W ($m^2 kg\, s^{-3}$)
pressure	pascal	Pa ($kg\, m^{-1} s^{-2}$)
solid angle	steradian	sr
temperature	kelvin	K
time	second	s
voltage	volt	V ($m^2 kg\, s^{-3} A^{-1}$)

Other Units, Numbers and Symbols in Common Use in Astronomy

Unit	Symbol	Equivalent
aeon		1,000,000,000 years
ångstrom	Å	10^{-10} m
astronomical unit	AU	$1.49597870 \times 10^{11}$ m
billion		1,000,000,000
degree	°	1/360 of a circle (≈ 0.0174563 rad)
dyne	dyn	10^{-5} N
electron-volt	eV	1.6022×10^{-19} J
erg	erg	10^{-7} J
gauss	G	10^{-4} T
jansky	jy	10^{-26} W m^{-2} Hz^{-1} (an obsolete synonym is flux unit (fu))
light year	ly	9.4605×10^{15} m (0.307 pc)
megasun	$MM_{\odot}$	A million solar masses = 2×10^{36} kg – a new term introduced by the author within this book
micron	μ, μm	10^{-6} m
million (mega) light years	Mly	9.4605×10^{21} m
minute of arc	′	1/60 th of a degree
parsec	pc	3.0857×10^{16} m (3.26 ly)
million (mega) parsecs	Mpc	3.0857×10^{22} m
second of arc	″	1/60 th of a minute of arc (1/3,600 of a degree)
milliarcsecond	mas	1/1,000 of a second of arc (1/3,600,000 of a degree)
solar luminosity	$L_{\odot}$	3.8478×10^{26} W
solar mass	$M_{\odot}$	1.9891×10^{30} kg
solar radius	$R_{\odot}$	6.960×10^{8} m
trillion		1,000,000,000,000
velocity of light in a vacuum	c	2.997925×10^{8} m s^{-1} (300,000 km/s is close enough most of the time)
year	yr	31,556,926 seconds of time (31.5 million or even 30 million is close enough most of the time)

Table A5.1 Interconversion of AGN bolometric luminosity measures.

Absolute bolometric magnitude	Luminosity (W)	Luminosity ($L_\odot$)	Luminosity (×bolometric luminosity of the Milky Way Galaxy)	Comments
−5	2.5×10^{30}	6.3×10^{3}	1.8×10^{-7}	
−6	6.3×10^{30}	1.6×10^{4}	4.5×10^{-7}	
−7	1.6×10^{31}	4.0×10^{4}	1.1×10^{-6}	
−8	4.0×10^{31}	1.0×10^{5}	2.9×10^{-6}	Faint normal elliptical galaxy
−9	1.0×10^{32}	2.5×10^{5}	7.2×10^{-6}	Brightest individual stars
−10	2.5×10^{32}	6.3×10^{5}	1.8×10^{-5}	Faintest detectable AGNs
−11	6.3×10^{32}	1.6×10^{6}	4.5×10^{-5}	
−12	1.6×10^{33}	4.0×10^{6}	1.1×10^{-4}	
−13	4.0×10^{33}	1.0×10^{7}	2.9×10^{-4}	
−14	1.0×10^{34}	2.5×10^{7}	7.2×10^{-4}	
−15	2.5×10^{34}	6.3×10^{7}	1.8×10^{-3}	Faint normal spiral galaxy
−16	6.3×10^{34}	1.6×10^{8}	4.5×10^{-3}	
−17	1.6×10^{35}	4.0×10^{8}	0.011	
−18	4.0×10^{35}	1.0×10^{9}	0.029	
−19	1.0×10^{36}	2.5×10^{9}	0.072	

−20	2.5×10^{36}	6.3×10^{9}	0.18	
−21	6.3×10^{36}	1.6×10^{10}	0.45	Type-I supernovae at maximum
−22	1.6×10^{37}	4.0×10^{10}	1.1	~Milky Way Galaxy values
−23	4.0×10^{37}	1.0×10^{11}	2.9	Seyfert/QSO borderline and
−24	1.0×10^{38}	2.5×10^{11}	7.2	brightest normal galaxies
−25	2.5×10^{38}	6.3×10^{11}	18	
−26	6.3×10^{38}	1.6×10^{12}	45	3C273
−27	1.6×10^{39}	4.0×10^{12}	110	
−28	4.0×10^{39}	1.0×10^{13}	290	
−29	1.0×10^{40}	2.5×10^{13}	720	
−30	2.5×10^{40}	6.3×10^{13}	1.8×10^{3}	Brightest QSOs and quasars
−31	6.3×10^{40}	1.6×10^{14}	4.5×10^{3}	
−32	1.6×10^{41}	4.0×10^{14}	1.1×10^{4}	~Brightest ULIRGs (perhaps)
−33	4.0×10^{41}	1.0×10^{15}	2.9×10^{4}	
−34	1.0×10^{42}	2.5×10^{15}	7.2×10^{4}	

Appendix 5
Greek Alphabet

Letter	Lower case	Upper case
Alpha	α	A
Beta	β	B
Gamma	γ	Γ
Delta	δ	Δ
Epsilon	ε	E
Zeta	ζ	Z
Eta	η	H
Theta	θ	Θ
Iota	ι	I
Kappa	κ	K

Letter	Lower case	Upper case
Lambda	λ	Λ
Mu	μ	M
Nu	ν	N
Xi	ξ	Ξ
Omicron	ο	O
Pi	π	Π
Rho	ρ	P
Sigma	σ	Σ
Tau	τ	T
Upsilon	υ	Y
Phi	φ	Φ
Chi	χ	X
Psi	ψ	Ψ
Omega	ω	Ω

Index

Page numbers that are underlined indicate the start of a major section on the topic.

Index

Index

Printed in Singapore